ZigBee Wireless Sensor and Control Network

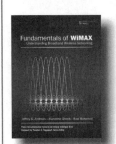

ZigBee Wireless Sensor and Control Network

Ata Elahi, Ph.D.
with Adam Gschwender

PRENTICE
HALL

Upper Saddle River, NJ • Boston • Indianapolis • San Francisco
New York • Toronto • Montreal • London • Munich • Paris • Madrid
Cape Town • Sydney • Tokyo • Singapore • Mexico City

The publisher offers excellent discounts on this book when ordered in quantity for bulk purchases or special sales, which may include electronic versions and/or custom covers and content particular to your business, training goals, marketing focus, and branding interests. For more information, please contact:

U.S. Corporate and Government Sales
(800) 382-3419
corpsales@pearsontechgroup.com

For sales outside the United States please contact:

International Sales
international@pearson.com

Visit us on the Web: informit.com/ph

Library of Congress Cataloging-in-Publication Data:

Elahi, Ata.
   ZigBee wireless sensor and control network / Ata Elahi, with Adam Gschwender
      p. cm.
      ISBN 978-0-13-713485-4 (pbk. : alk. paper) 1. Personal communication service systems—Standards. 2. Sensor networks. 3. Low voltage systems—Industrial applications. 4. Radio frequency—Industrial applications. I. Gschwender, Adam, 1979- II. Title.
   TK5103.485.E43 2009
   004.6'2—dc22   2009032393

ISBN-13: 978-0-137-13485-4
ISBN-10: 0-137-13485-1

Text printed in the United States on recycled paper at RR Donnelley & Sons, Crawfordsville, Indiana
First printing November 2009

In memory of my mother.

—Ata Elahi

To my parents for all their love and support.

—Adam Gschwender

CONTENTS

PREFACE

ZigBee is a new personal-area network (PAN) standard developed by the ZigBee Alliance. The ZigBee standard can be used to establish a wireless PAN, specifically a low-rate/power wireless sensor and control network. Wireless sensor and control networking is one the most rapidly growing technologies and has a wide variety of applications, including smart energy; commercial building automation; home automation; personal, home, and hospital care; remote-control applications for consumer electronics; telecom applications; and wireless sensor network applications.

This book presents an overview of the ZigBee technology and its applications, allowing the wireless system designer, manager, or student access to this new and growing field of wireless sensor and control networking. For the uninitiated in wireless technology, the book provides a helpful overview of wireless technology, giving the reader the background necessary for understanding ZigBee. It goes into detail about the ZigBee protocol stack, describing ZigBee's use of IEEE 802.15.4, which defines the Media Access Control (MAC) and physical layers for the low-rate wireless personal-area network (LR-WPAN), and ZigBee's implementations of the network, security, and application layers.

ORGANIZATION

This book is divided into 11 chapters, which commence by introducing you to wireless technology and then proceed up the ZigBee protocol stack. In aggregate, the chapters provide comprehensive coverage of IEEE 802.15.4 and the ZigBee protocol architecture. In addition, three appendixes describe alternative technologies that can also be used to establish a PAN.

Chapter 1, "Introduction to Wireless Networks," covers the Open Systems Interconnection (OSI) reference model; error detection; the Industrial, Scientific, and Medical (ISM) band; modulation techniques; wireless local-area networks (WLANs), frequency-hopping spread spectrum (FHSS); direct-sequence spread spectrum (DSSS); wireless metro-area networks (MANs); and Bluetooth.

Chapter 2, "ZigBee Wireless Sensor and Control Network," presents an overview of ZigBee applications, ZigBee characteristics, ZigBee device types, ZigBee topologies, ZigBee protocol architecture, and characteristics of ZigBee PRO.

Chapter 3, "IEEE 802.15.4 Physical Layer," covers frequency bands, data rate, channels, the physical layer data and management services, transmitter power, receiver sensitivity, received signal strength indication (RSSI), and link-quality indication.

Chapter 4, "IEEE 802.15.4 Media Access Control (MAC) Layer," covers MAC data and management services, the MAC layer information base, access methods, the beacon frame, the MAC data frame and control frame, and the command frame format.

Chapter 5, "Network Layer," covers the network layer data entity; the Network Information Base (NIB); the configuration of a new device; starting a network; addressing, joining, and leaving a network; network discovery; channel scanning; the network-formation process; route discovery; and the network command frame format.

Chapter 6, "Application Support Sublayer (APS)," covers the application support sublayer data and management entities, the APS Information Base, the APS sublayer frame format, and the APS command frame format.

Chapter 7, "Application Layer," presents the application profile, attribute, cluster, cluster format, general cluster commands, ZigBee cluster libraries, simple application profile, ZigBee device profile, node descriptor, and binding and network management commands.

Chapter 8, "Security," covers elements of network security, Advanced Encryption Standard (AES), ZigBee security and the Trust Center, ZigBee Residential, Standard and High-Security modes, ZigBee security management primitives, counter mode encryption (CTR), and cipher block chaining encryption (CBC).

Chapter 9, "Address Assignment and Routing," covers address assignment using distributed schemes, stochastic address assignment, Ad hoc On-Demand Distance Vector (AODV) Routing protocol, unicast routing discovery, multicast routing discovery, dynamic source routing, ZigBee routing attributes, tree hierarchical routing, ZigBee PRO routing, and routing commands.

Chapter 10, "ZigBee Home Automation and Smart Energy Network," examines the ZigBee home automation cluster, home automation network requirements, devices used for home automation, commissioning, the Smart Energy network, advanced metering infrastructure (AMI), and home-area networks (HANs).

Chapter 11, "ZigBee RF4CE," covers the Radio Frequency for Consumer Electronics (RF4CE) protocol, RF4CE nodes and topology, network layer data and management services, and the pairing process.

Appendix A, "6lowpan," covers IPv6 over low-power wireless personal-area network (6LoWPAN).

Appendix B, "Wireless HART," covers wireless HART.

Appendix C, "Z-Wave," covers Z-Wave technology.

ACKNOWLEDGMENTS

Many people contributed to the development of this book. We want to express our deep appreciation to Spiro Sacre of National Technical System for his in-depth review of the manuscript and his valuable suggestions and comments, which enabled us to improve the quality of this book. We also want to thank the following reviewers who reviewed the manuscript and provided valuable suggestions for its improvement: Ryan J. Maley, vice president of operations at Software Technologies Group; and Ian Marsden, director of Integration Associates, and Dr. Farid Farahmand, assistant professor, Sonoma State University.

And for their encouragement and support, we also want to thank Dr. Edward Harris, Dean of the School of Communication, Information and Library Science; Professor Winnie Yu, chairperson of Computer Science at Southern Connecticut State University; and Reza Khani, vice president of operations, Petra Solar Inc. And a special thanks to the staff of Pearson, especially Bernard Goodwin, Lori Lyons, Keith Cline, and Michelle Housley.

ABOUT THE AUTHORS

Ata Elahi has been a professor in the Computer Science Department of Southern Connecticut State University since 1986. His research areas include computer networks, data communication, computer hardware design, and pipeline processors. Elahi's books include *Data, Network, and Internet Communications Technology* and *Communication Network Technology*. He holds a Ph.D. in electrical engineering from Mississippi State University.

Adam Gschwender, a professional software engineer with wide-ranging experience, currently develops advanced search-related applications.

CHAPTER 1

INTRODUCTION TO WIRELESS NETWORKS

INTRODUCTION

Wireless sensor and control networks are quickly becoming an integral part of the automation process within chemical plants, refineries, and commercial buildings. The market for wireless sensor and control networks is growing fast; so much so that, according to the *RIFD Journal* (www.rfidjournal.com/article/3634/-1/1), it will be a $5 billion industry by the year 2011.

To accommodate this burgeoning technology, numerous standards are being developed for wireless sensor and control networking, including the following:

- SP100.11 (Wireless Systems for Automation) by the Industrial Standard for Automation (ISA)
- Wireless HART (Highway Addressable Remote Transducer) by the HART organization
- IPv6 over low-power personal-area network (6lowpan) by IETF (the Internet Engineering Task Force)
- ZigBee by the ZigBee Alliance

Moreover, the multiple standards for wireless technologies that currently exist can be used for data transfer: WLAN (Wireless LAN), Bluetooth, Wireless MAN (IEEE 802.16), and Ultra Wideband (IEEE 802.15.3). With so many standards vying for a network engineer's attention, it is critical, at the very least, to have a general understanding of each. However, before we go into greater depth on any one standard, the general application of a wireless sensor and control network must be described. The following is a list of the more common applications for wireless sensor and control networks:

- **Building and home automation**

  Door and garage control

  Automatic meter reading

  Lighting control

  Security monitoring

- **Industrial and process automation**

  Temperature sensing and control

  Pressure sensing

  Flow control

  Level sensing

  Monitoring air quality

- **Energy and utility automation**

  Power distribution

  Meter monitoring

- **RFID and logistics**

  Container security

  Cold chain monitoring (The McDonald's Corporation requires that each restaurant record its freezer temperature several times per day and the freezer temperature of the truck that delivers frozen food to the restaurant.)

- **Miscellaneous monitoring**

  For example, a football player's helmet may be equipped with wireless sensors which, when a player receives a hard impact to his helmet, will transmit the force of the impact to the coach so that he may decide whether the player should continue to play.

Traditionally, a machine's condition has been manually monitored so as to prevent sudden failures. More recently, this has been a task delegated to wireless sensor networks (WSNs)—those networks that can monitor a machine's pressure and temperature, for example. Several companies currently produce WSN products for monitoring the condition of a machine: Coronis System, Dust Networks, Honeywell, and Sensicast. Most installed WSNs use proprietary technology rather than a standard technology. However, there has been a general push by manufacturers of wireless sensor and control networks for standardization. For those standards that have been developed, most use IEEE 802.15.4 for the physical and data link layers.

The IEEE 802.15 working group has developed three types of standards for wireless personal-area networks (WPANs):

1. IEEE 802.15.1 is the standard for Bluetooth, which is generally used as a cable replacement for computer peripherals (for example, a mouse/keyboard).

2. IEEE 802.15.3 is used for multimedia applications that require a high quality of service.

3. IEEE 802.15.4 is the standard for low-rate wireless personal-area network (LR-WPAN). LR-WPAN is used for applications that require low data rates and consume less power.

1.1 THE OPEN SYSTEMS INTERCONNECTION (OSI) REFERENCE MODEL

The Open Systems Interconnection (OSI) reference model was developed by the International Organization for Standardization (ISO) for interoperability between equipment designed for networks. An open system is a set of protocols that allow two computers to communicate with each other regardless of their design, manufacturer, or CPU type. Any device that obeys the OSI standard can be easily connected to any other device that also adheres to the standard. The OSI model divides network communications into seven layers, with each layer performing the specific tasks, as shown in Figure 1.1.

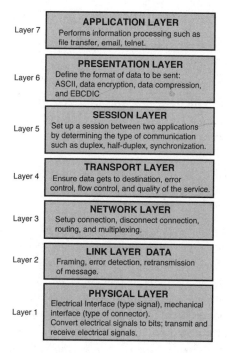

Figure 1.1 The OSI reference model

1.1.1 Layer 1: Physical Layer

The physical layer defines the type of signal and connectors (for example, RS-232 or RJ-45) to be used by the network interface card (NIC). It also defines the cable types (coaxial cable, twisted-pair, or fiber-optic cable) used as the transmission medium. The physical layer accepts and sends signal transmissions and performs the signal-to-bit and bit-to-signal conversions. Within wireless devices, the physical layer uses radio frequency (RF) for the transmission of information. The physical layer also performs modulation and demodulation.

1.1.2 Layer 2: Data Link Layer

The data link layer defines the frame format, which includes the start and end of the frame, frame size, and other frame specifics. Specifically, it performs the following functions:

- **On the transmitting side:** The data link layer accepts information from the network layer and breaks the information into frames. It then adds the destination Media Access Control (MAC) address, source MAC address, and the Frame Check Sequence (FCS) field, and finally passes each frame to the physical layer for transmission.
- **On the receiving side:** The data link layer accepts bits from the physical layer and forms them into a frame, performing error detection. If the frame is free of errors, the data link layer passes the frame up to the network layer.
- **Frame synchronization:** It identifies the beginning and end of each frame.
- **Flow control:** Controls rate of transmission; the transmission rate should not be higher than the processing rate of the receiver station.
- **Link management:** It coordinates transmission between transmitter and receiver.
- **Determine contention method:** It defines an access method in which two or more network devices compete for permission to transmit information across the same communication media, such as token passing or carrier-sense multiple access with collision detection (CSMA/CD).

1.1.3 Layer 3: Network Layer

The function of the network layer is to perform routing. Routing determines the route or pathway for moving information over a network (in a network with multiple local-area networks [LANs]). The network layer determines the logical address of each frame and then forwards that frame to the next router indicated in its routing table. It is responsible for translating each logical address (name address) to a physical address (MAC address). An example of a network layer protocol is the Internet Protocol (IP).

The network layer provides two types of services: connectionless and connection-oriented services. In connection-oriented services, the network layer makes a connection between source and destination and then starts the transmission. In connectionless services, there is no connection between the source and destination; the source transmits information regardless of whether the destination is ready. A common example of this type of service is email.

1.1.4 Layer 4: Transport Layer

The transport layer provides reliable transmission of data to ensure that each frame reaches its destination. If, after a certain period of time, the transport layer does not receive an acknowledgment from the destination, it retransmits the frame and again waits for an acknowledgment. An example of a transport layer protocol is the Transmission Control Protocol (TCP).

1.1.5 Layer 5: Session Layer

The session layer establishes a logical connection between the applications of two computers that are communicating with each other. It allows two applications on two different computers to establish and terminate a session. For example, when a workstation connects to a server, the server performs the login process, requesting a username and password, and, upon successful authentication, establishes a session.

1.1.6 Layer 6: Presentation Layer

The presentation layer receives information from the application layer and converts it to a form acceptable to the destination. The presentation layer can convert information to ASCII or Unicode, or encrypt or decrypt the information.

1.1.7 Layer 7: Application Layer

The application layer enables users to access the network with applications such as email, File Transfer Protocol (FTP), and Telnet.

1.2 IEEE 802 STANDARD COMMITTEE

The Institute of Electrical and Electronics Engineers (IEEE) 802 committee originally defined the standards for the physical layer and the data link layer in February 1980, calling it IEEE 802, with 80 representing 1980, and 2 representing the month of February. Figure 1.2 shows the difference between the IEEE 802 standard and OSI model. The IEEE standard divides the data link layer of the OSI model into two sublayers: Logical Link Control (LLC) and Media Access Control (MAC).

- **MAC:** The MAC layer defines the method that a node uses to access the network:
  - Carrier-sense multiple access with collision detection (CSMA/CD) is used for Ethernet.
  - A control token is used in Token Ring networks and Token Bus networks.

- Carrier-sense multiple access with collision avoidance (CSMA/CA) is used for wireless networks.
- **LLC:** The LLC defines the format of the frame. It is independent of a network's topology, transmission media, and MAC.

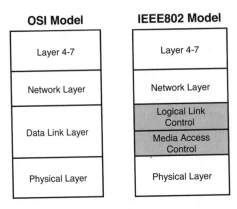

Figure 1.2   The OSI and IEEE standard model

Figure 1.3 shows the different MAC layers for several IEEE 802 networks. All networks that are listed use the same logical link control. IEEE 802.11 is the standard for Wireless LAN, and IEEE 802.16 is the standard for Wireless MAN.

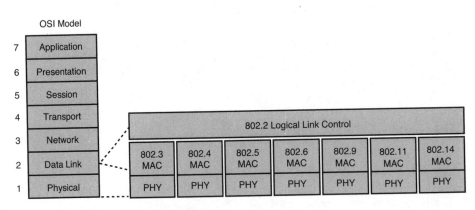

Figure 1.3   IEEE 802 reference model

1.3 WIRELESS TECHNOLOGIES

Two common types of technologies are used for transmission of information in wireless devices:

- **Infrared (IR) technology:** IR technology is most suitable for indoor use because infrared rays cannot penetrate walls, ceilings, or other obstacles. That is, the transmitter and receiver must have a line of sight between each other, just like the remote control for a television set. In an environment where there are obstacles such as buildings and walls between the transmitter and a receiver, the transmitter may use diffused IR. However, most wireless devices use RF technology.

- **Radio frequency (RF) technology:** There are two types of RF signals used for transmission of information: narrowband signal and spread-spectrum signal.

  - *Narrowband signal:* The narrowband signal refers to a signal with a narrow spectrum, as shown in Figure 1.4. In narrowband, the information is transmitted at a specific frequency, such as those used over AM or FM radio waves.

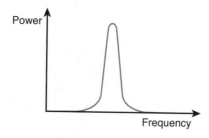

Figure 1.4   Narrowband signal

  - *Spread-spectrum signal:* In spread-spectrum technology, the information is transmitted over a range of frequencies, as shown in Figure 1.5. Spread-spectrum is one of the most popular signal types for wireless devices.

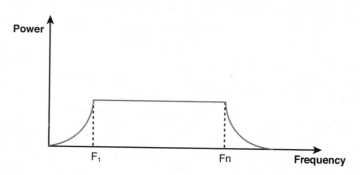

Figure 1.5   Spread-spectrum signal

There are certain advantages to using the spread-spectrum band over the narrowband, including the following:

- In spread-spectrum technology, information is transmitted at different frequencies.
- It is difficult to jam a spread-spectrum signal; the signal cannot be easily disrupted by other signals.

- Interception of spread-spectrum signals is more difficult than interception of a narrow-band signal.
- Noise is less disruptive in spread-spectrum signals than in a narrowband signal.

1.4 ANTENNA

An antenna is a conductor that is used to radiate and receive electromagnetic waves and is characterized by its directionality and gain:

- **Directionality:** This refers to the direction in which the RF signal is transmitted by the antenna. The two types of directionality are omnidirectional, transmitting the RF signal 360 degrees around the antenna, and directional, transmitting in a specific direction. Figure 1.6 shows both directionality types.

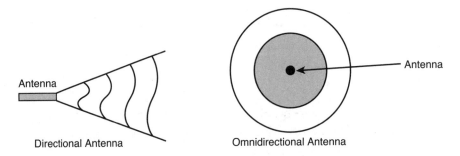

Antenna

Antenna

Directional Antenna

Omnidirectional Antenna

Figure 1.6 Directional and omnidirectional antennas

- **Gain:** This is measured in dBi, where dB stands for decibel, and i stands for isotropic. An isotropic antenna is an ideal antenna that transmits the RF signal in all directions equally. However, practical antennas do not transmit RF signals equally in all directions. The gain of the antenna is given by Equation 1.1.

$$G = P_a / P_i \tag{1.1}$$

Where:

G: Antenna gain

P_i: Power density of isotropic antenna at the same distance as defined below

P_a: Power density of real antenna in specific direction and distance

$P_i = P_t / 4 \pi r^2$

P_t: Transmitted power in watts

r: The distance from the antenna in meters

1.5 ERROR-DETECTION METHODS

When the transmitter sends a frame to the receiver, the frame can become corrupted due to external and internal noise, which requires the receiver to first check the integrity of the frame. Possible sources of error include the following:

- **Impulse Noise:** A noncontinuous pulse for a short duration is called impulse noise. It may be caused by a lightning discharge or a spike generated by a power switch being turned off or on.

- **Attenuation:** When a signal propagates, the strength of the signal is reduced over distance. This reduction is called attenuation. A weak signal is more affected by noise than a strong signal.

- **White noise or thermal noise:** This type of noise exists in all electrical devices and is generated by moving electrons in the conductor.

- **Radio interference:** This type of noise is caused by other wireless transmitters using the same channel.

The following methods can be used to detect an error or errors:

- **Parity check:** The simplest error detection method is the parity check. The parity check method can detect one error.

- **Block check character (BCC):** Uses vertical and horizontal parity bits to detect double errors.

- **One's complement of the sum:** The method used for error detection in the TCP header and IP header. At the transmitter side, the 16-bit one's complement sum of the header is calculated. The result of this calculation is transmitted with the information to the receiver. At the receiver side, the 16-bit one's complement of the header is calculated and compared to the result with the one's complement of the transmitter. If the two results are equal, no error is detected. Otherwise, there is an error in the information.

- **Cyclic redundancy check (CRC):** This method is used to detect one or more errors.

1.5.1 Cyclic Redundancy Check (CRC)

The parity bit and BCC can detect single and double errors. The CRC method is used for detection of a single error, more than a single error, and a burst error (when two or more consecutive bits in a frame have changed).

The CRC uses modulo-2 addition to compute the frame check sequence (FCS). In modulo-2, addition

$$1 + 1 = 0, 1 + 0 = 1, \text{ and } 0 + 0 = 0$$

The following procedure is used to calculate FCS:

- **Transmitter side:** Frame M is k bits, P is a divisor of n + 1 bits, FCS is n bits, and it is the remainder of 2^n* M/P using modulo-2 division. After these values have been calculated, the transmitter will transmit the frame, T = 2^n* M + FCS, to the receiver, where T is k + n bits.
- **Receiving side:** The receiver divides T by P using modulo-2 division. If the result of this division generates a remainder of zero, no error is detected in the frame. Otherwise, the frame contains one or more errors.

Example

Find the FCS for the following message. The divisor is given.
Message M = 111010, K = 6 bits
Divisor P = 1101, n + 1 = 4 bits
Therefore 2^n * M = 111010000
By dividing 2^n * M by P using modulo-2 division, FCS = 010, as shown in Figure 1.7.

```
              101010
        1101 |111010000
              1101
              1110
              1101
               1100
               1101
                010  Remainder or FCS
```

Figure 1.7 FCS calculations

At the transmitter side, the FCS is added to 2^n * M, and the transmitter transmits frame T = 111010010 to the receiver.

At the receiver side, the receiver divides T by P and, if the result has a remainder of zero, there is no error in the frame. Otherwise, the message contains an error. Because the above division takes time, special hardware is designed to generate the FCS.

CRC polynomial and architecture: A binary number, $b_5b_4b_3b_2b_1b_0$, where each bit is represented in the polynomial:

$$b_5X^5 + b_4 X^4 + b_3 X^3 + b_2X^2 + b_1X + b_0$$

Example

Represent P = 1101101 as a polynomial.

$$P(X) = 1 * X^6 + 1 * X^5 + 0 * X^4 + 1 * X^3 + 1 * X^2 + 0 * X + 1 = X^6 + X^5 + X^3 + X^2 + 1$$

The following CRC polynomials are IEEE and ITU standards:

$$CRC\text{-}12 = X^{12} + X^{11} + X^3 + X^1 + 1$$
$$CRC\text{-}16 = X^{16} + X^{15} + X^2 + 1$$
$$CRC\text{-}ITU = X^{16} + X^{12} + X^5 + 1$$
$$CRC\text{-}32 = X^{32} + X^{26} + X^{23} + X^{22} + X^{16} + X^{11} + X^{10} + X^8 + X^7 + X^5 + X^4 + X^2 + X + 1$$

The CRC-16 is used by the IEEE 802.15.4 MAC layer.
In general, a CRC polynomial can be represented by

$$P(X) = X^n + \dots + a_4 X^4 + a_3 X^3 + a_2 X^2 + a_1 X + 1$$

Figure 1.8 shows the general architecture of a CRC integrated circuit (IC). C_i is a 1-bit shift register, and the output of each register is connected to the input of an Exclusive-OR gate; a_i is the coefficient of a CRC polynomial. In Figure 1.8, if a_1 equals zero, then there is no connection between the feedback line and the XOR gate. To find the FCS, the initial value for C_i is set to zero and the message $2^n * M$ is shifted k + n times through the CRC circuit. The final contents of $C_{n-1}, \dots C_4, C_3, C_2, C_1, C_0$ is the FCS.

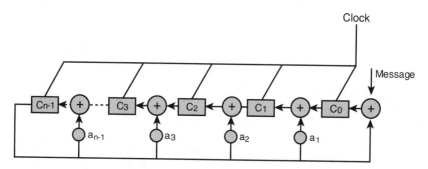

Figure 1.8   General architecture of a CRC polynomial

Example

Show the CRC circuit for a polynomial:

$$P(X) = X^5 + X^4 + X^2 + 1$$

In the above polynomial, the value for a_1, a_3 are zero. Figure 1.9 shows the CRC circuit for the above polynomial.

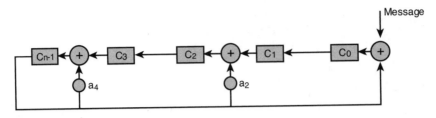

Figure 1.9   CRC circuit for polynomial $P(X) = X^5 + X^4 + X^2 + 1$

Example

Find FCS message M = 111010, assume P = 1101

$$P(X) = X^3 + X^2 + 1$$

The circuit for P(X) is shown in Figure 1.10, where $a_1 = 0$, $a_2 = 1$, and $a_3 = 1$.

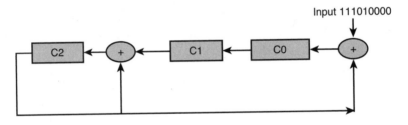

Figure 1.10   CRC circuit for P = 1101

Table 1.1 shows the contents of each register after shifting 1 bit at a time. After shifting 9 (k + n) times, the content of the registers is the FCS.

Table 1.1   FCS Value of 010 for Message M = 111010 and P = 1101

C2	C1	C0	Input
0	0	0	Initial value
0	0	1	1
0	1	1	1
1	1	1	1
0	1	1	0
1	1	1	1
0	1	1	0
1	1	0	0
0	0	1	0
0	1	0	0

1.6 ISM AND U-NII BANDS

- **Industrial, Scientific, and Medical (ISM) band:** The Federal Communication Commission (FCC) allocates a separate range of frequencies to radio stations, TV stations, telephone companies, and navigation and military agencies. The FCC also allocates a band of frequencies called the ISM band for industrial, research, and medical applications. The use of the ISM band does not require a license from the FCC for transmissions consuming up to 1 watt of power. Figure 1.11 shows the frequency allocations for the ISM band.

902 MHz	928 MHz	2.4 GHz	2.48 GHz	5.725 GHz	5.85 GHz
Industrial Band I-band		Scientific Band S-band		Medical Band M-band	

Figure 1.11 ISM band

- **Unlicensed National Information Infrastructure (U-NII) band:** The U-NII band consists of three 100MHz frequency bands, where each band uses a specific transmission power, as shown in Figure 1.12.

5.15GHz	5.25 GHz	5.35GHz	5.725GHz	5.82GHz
40 mW	200 mW		800 mW	

Figure 1.12 U-NII band

1.7 MODULATION

Within wireless devices, one of the functions of the physical layer is to convert the digital signal into an analog signal for transmission. This process is known as modulation and can be performed using several techniques, such as following methods.

- **Amplitude-shift keying (ASK):** This method uses changes of amplitude to represent zero and one. As shown in Figure 1.13, the smaller amplitude represents *zero,* and the larger amplitude represents *one.* Each cycle represents 1 bit; therefore, in this case, the baud rate is equivalent to the number of bits per second.
- **Frequency-shift keying (FSK):** A *zero* is represented by no change in the frequency of the original signal, while a *one* is represented by a change to the frequency of the original signal, as shown in Figure 1.14.

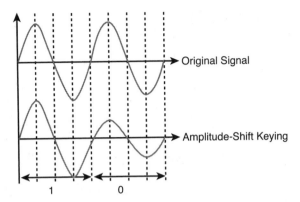

Figure 1.13   Amplitude-shift keying (ASK)

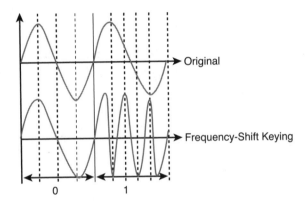

Figure 1.14   Frequency-shift keying (FSK)

- **Phase-shift keying (PSK):** In this modulation technique, the phase of the signal is used to represent the binary data. To illustrate, Figure 1.15 shows a 90-degree phase shift. Figure 1.16 (a), (b), and (c) show the original signals with a 90-degree shift, a 180-degree shift, and a 270-degree shift, respectively. As Figure 1.15 and Figure 1.16 indicate, the original signal can be represented with four different signals: no shift, a 90-degree shift, a 180-degree shift, and a 270-degree shift. Therefore, each cycle can represent a 2-bit binary number by employing one of the four phase-shifted signals, as shown by Table 1.2.

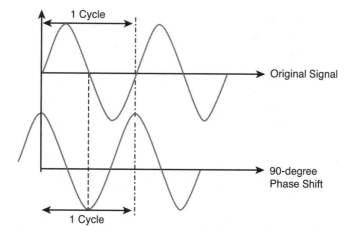

Figure 1.15   90-degree phase shift

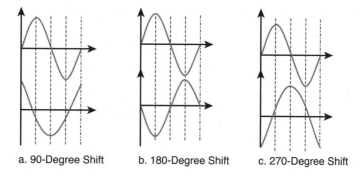

a. 90-Degree Shift   b. 180-Degree Shift   c. 270-Degree Shift

Figure 1.16   Phase shift for 90, 180, and 270 degrees

Table 1.2   Phase Shifts and Their Binary Representation

Phase Shift	Binary Value
No shift	00
90°	01
180°	10
270°	11

There are several types of phase shift keying:

Binary phase-shift keying (BPSK): The signal is shifted by 180 degrees to represent binary 1, whereas no shift represents binary 0.

Quadrature phase-shift keying (QPSK) or 4-PSK: Each signal is shifted by increments of 90 degrees. Table 1.2 shows the 90-degree phase shifts and their corresponding binary values.

8-PSK: The signal is shifted by increments of 45 degrees, allowing for eight different phase shifts. Each cycle, then, can represent a 3-bit binary number, as shown by Table 1.3.

Table 1.3   8-Phase-Shift Keying (8-PSK)

Bits	Phase Shift
000	0
001	45
010	90
011	135
100	180
101	225
110	270
111	315

- **Quadrature amplitude modulation (QAM):** QAM is the combination of PSK and amplitude modulation. As shown in Figure 1.17, the combination of four phases and two amplitudes generates eight different signals, which together are known as 8-QAM. Table 1.4 shows the binary value of each signal

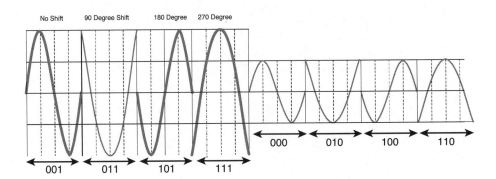

Figure 1.17   8-QAM modulation

Table 1.4   Binary Values for 8-QAM

Shift	Amplitude	Binary Value
No	A1	000
No	A2	001
90°	A1	010
90°	A2	011
180°	A1	100
180°	A2	101
270°	A1	110
270°	A2	111

1.8   WIRELESS LOCAL-AREA NETWORK (WLAN)

The WLAN or IEEE 802.11 enables users to access an organization's network from any location inside the organization without any physical connection to the organization's network. WLAN uses FR or IR waves as its transmission media. WLAN is staged to be the next generation of campus networking.

The IEEE 802.11 committee has approved several standards for WLAN that define the functions for the MAC and physical layers. Table 1.5 shows the physical layer and data link layer for various WLAN standards.

Table 1.5   IEEE 802.11 Physical Layer and Data Link Layer

Logical Link Control					Data Link Layer
Medium Access Control					
IEEE 802.11 DSSS and FHSS 1 and 2 Mbps	IEEE 802.11b HR-DSSS 1, 2, 5.5, and 11 Mbps	IEEE 802.11a OFDM 6, 12, 18, 24, 36 45 and 54 Mbps	IEEE 802.11g OFDM 6, 12, 18, 24, 36, 45 and 54 Mbps	IEEE 802.11n OFDM MIMO 600 Mbps	Physical Layer
ISM band	ISM band	U-NII band	ISM band	ISM	

1.8.1 Wireless LAN Physical Layer

The wireless physical layer performs the following functions:

- Modulation and encoding. Information is modulated and then transmitted to the destination.
- Supports multiple data rate.
- Senses the channel to see whether it is clear (carrier sense).
- Transmits and receives information.

1.8.2 Physical Layer Standards

- **IEEE 802.11 physical layer:** The IEEE 802.11 standard operates in the ISM band and is designed for data rates of 1Mbps and 2Mbps. It supports two types of radio frequencies for data transmission: frequency-hopping spread spectrum (FHSS) and direct-sequence spread spectrum (DSSS).

- **IEEE 802.11b physical layer:** The IEEE 802.11b standard extends the DSSS physical layer of 802.11 to provide higher data rates of 5.5Mbps and 11Mbps. 802.11b uses complementary code keying (CCK) to support the two new data rates, 5.5Mbps and 11Mbps, in addition to 1Mbps and 2Mbps. IEEE 802.11b data rates are shown in Table 1.6.

Table 1.6 IEEE 802.11 Data Rates and Modulations

Data Rate Mbps	Code Length	Modulation Method	Signal Rate	Bits/Signal
1	11 chip bits	BPSK	1Mbps	1
2	11 chip bits	QPSK	1Mbps	2
5.5	8 CCK	QPSK	1.375M	4
11	8 CCK	QPSK	1.375M	8

The IEEE 802.11b standard defines 11 channels that may be used. Each channel is represented by its center frequency, which is shown in Table 1.7. As indicated by the table, each channel is separated from adjacent channels by 5MHz. However, because the bandwidth of each channel is 16MHz, using adjacent channels will cause interference. IEEE 802.11b supports three nonoverlapping channels (1, 6, and 11) to overcome interference problems.

Table 1.7 IEEE 802.11b Channels and Frequencies

Channel Number	Center Frequency (MHz)
11	2462
10	2457
9	2452
8	2447
7	2442
6	2437
5	2432
4	2427
3	2422
2	2417
1	2412

- **IEEE 802.11g physical layer:** IEEE.80211g operates at 2.4GHz using DSSS and ODFM for transmission of information.

- **IEEE 802.11n:** IEEE802.11n uses multiple-input multiple-output (MIMO) to receive and transmit information. MIMO uses multiple transmitter and receiver antennas to improve the data rate. This standard will be ratified by 2009, but some corporations have already begun producing wireless network components for IEEE802.11n.

1.8.3 WLAN Media Access Control (MAC) Layer

The MAC layer performs the following functions in a WLAN.

- Supports multiple physical layers
- Supports access control
- Fragmentation of the frame

- Frame encryption
- Roaming

IEEE 802.11 supports the distribution coordination function; CSMA/CA and the point coordination function (PCF) as methods for a station to access wireless LANs.

1.8.4 Carrier–Sense Multiple Access with Collision Avoidance (CSMA/CA)

Most wireless networks, such as ZigBee and WLAN, are using CSMA/CA as an access method. When a station wants to transmit a frame, it first listens for signals transmitted over the medium. If there is no traffic, it continues to wait for a span of time known as the short interframe space. And if there is still no traffic on medium, the station will start transmitting; otherwise, it has to wait for the medium to become clear. Figure 1.18 shows the CSMA/CA flowchart operation.

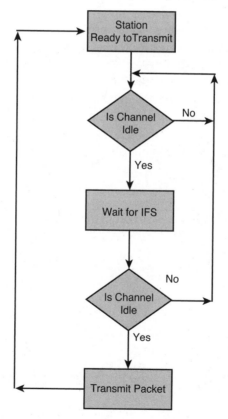

Figure 1.18   CSMA/CA flowchart

1.9 FREQUENCY-HOPPING SPREAD SPECTRUM (FHSS)

The IEEE 802.11 standard recommends using the scientific portion of the ISM band (2.4GHz to 2.483GHz) for WLAN. The FHSS divides the scientific band into 79 channels of 1MHz each. The transmitter divides the information and sends each part to a different channel, as shown in Figure 1.19. This process is known as frequency hopping.

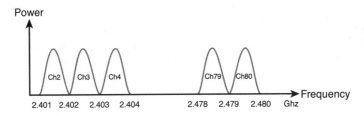

Figure 1.19   Frequency-hopping spread spectrum

The order of the channels or hop sequence used by the transmitter is predefined and has already been communicated to the receiver. For example, in a typical transmission, the hop sequence may be 3, 6, 5, 7, and 2. The hop sequence can be selected during the installation of the WLAN. The FCC requires that the dwell time, the time a transmitter spends in each frequency, be no greater than 400ms and uses 75 hop channels. The FCC also requires that the maximum power for the transmitter in the United States should not exceed 1 watt.

FHSS is more immune to noise because information is transmitted at different channels. If one channel is noisy, it can drop the noisy channel. Table 1.8 shows the FHSS channels and their frequencies.

Table 1.8   FHSS Channels and Frequencies

Channel Number	Frequency in GHz
2	2.402
3	2.403
4	2.404
.	.
.	.
.	.
80	2.480

1.10 DIRECT-SEQUENCE SPREAD SPECTRUM (DSSS)

In DSSS, each bit, before transmission, is broken down to a pattern of bits called a chip. The chip is generated by performing an XOR (Exclusive-OR) operation on each bit with a pseudo random code, as shown in Figure 1.20. The output of the XOR operation, the chip bit, is then modulated and transmitted. Figure 1.21 shows the transmission process performed by the physical layer. The receiver uses the same pseudo random code to decode the original data. IEEE 802.11 recommends 11 bits for each chip.

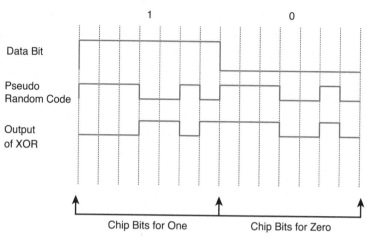

Figure 1.20   Generation chips

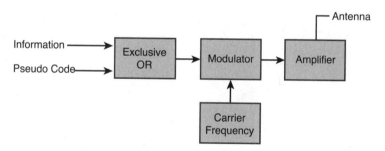

Figure 1.21   Physical layer using DSSS

The DSSS supports two types of modulation: differential binary phase-shift keying (DBPSK), which is used for a data rate of 1Mbps, and differential quadrature phase-shift keying (DQPSK), which is used for data rates of 2Mbps.

1.11 WIRELESS MAN

Wireless MAN, also known as Broadband Wireless Access (BWA), is a standard approved by the IEEE 802.16 committee in April 2002 for performing the following applications.

- Provides links between campus buildings and dormitories
- Provides links for hospitals within a city for sharing patient information

- Provides links between airport buildings and railways terminals
- Provides links for police and fire departments
- Provides wireless service to the cities

Wireless MANs can be used by ISPs to provide Internet access to buildings and offices, replacing cable and digital subscriber line (DSL) modems. Figure 1.22 shows an example of wireless MAN in which there are three types of stations: base station (BS), subscriber station (SS), and repeater. The base station is connected to the core network of an organization or a public network such as the Internet. The subscriber station serves users inside the building. As shown in the figure, all networks are connected to the base station without any cabling. The advantage of BWA is that it is easy to install, and it is less expensive than wired networks.

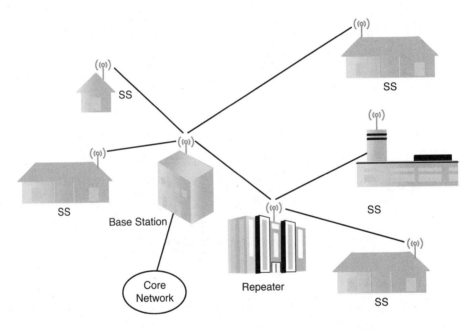

Figure 1.22   PMP wireless MAN topology

1.11.1 Wireless MAN Topology

IEEE 802.16 supports point to multipoint (PMP). In PMP topology, the base station transmits data to all subscriber stations in its cell, as shown in Figure 1.22. This transmission from the base station to the subscriber stations is called a down-link transmission. Similarly, a transmission from a subscriber station to the base station is called an uplink transmission. Figure 1.23 shows a mesh topology; mesh topologies are used for communication between subscriber stations using NLOS, which offers ad hoc networking between subscribers. Each subscriber station or mesh node can communicate with other nodes without requiring line of sight.

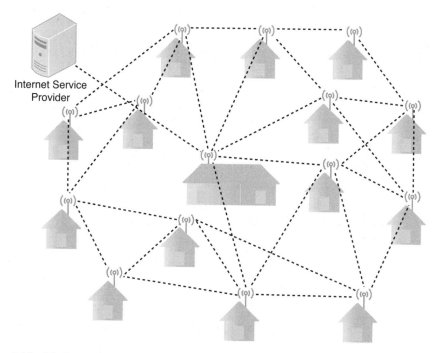

Figure 1.23 Mesh topology

1.11.2 IEEE 802.16 Protocol Architecture

IEEE 802.16 defines the MAC and physical layer standards for Wireless MAN. This standard includes three specifications for the physical layer: IEEE 802.16, IEEE 802.16a, and IEEE802.16e. The protocol architecture for Wireless MAN is shown in Figure 1.24.

Convergence Sublayer (CS)		MAC Layer
MAC Common Part Sublayer		
Security Sublayer		
Transmission Convergence		
IEEE802.16 10-66 GHz	IEEE802.16a 2-11Ghz	Physical Layer
QPSK 16QAM 64QAM	OFDM	

Figure 1.24 Protocol architecture of Wireless MAN

1.11.3 MAC layer

As shown in Figure 1.24, the MAC layer supports two physical layers. The MAC layer itself is divided into three sublayers:

- **Convergence sublayer (CS):** IEEE 802.16 is designed to carry various types of protocol data units (PDUs) such as ATM, IPv4, IPv6, and Ethernet. To achieve this, the convergence sublayer accepts the PDU information from the upper layer (logical link control) and converts it to the IEEE 802.16 MAC data format. It also performs the reverse, converting the 802.16 MAC data format to other protocol data units such as ATM or IP.
- **MAC common part sublayer (CPS):** The IEEE 802.16 MAC layer uses a connection-oriented PMP transmission. The base station uses an antenna with multiple sectors to transmit data to all subscriber stations simultaneously without any coordination.
- **Security sublayer:** The security sublayer performs authentication between the BS and SS encryption and decryption. The security sublayer uses X.509 certificates for authentication and 56-bit Data Encryption Standard (DES) for encryption.

1.11.4 Physical Layers

IEEE approved three physical layers for Wireless MAN: IEEE 802.16, IEE802.16a, and IEEE 802.16e. Table 1.9 shows the characteristics of IEEE 802.16, IEEE 802.16a, and IEEE 802.16e.

Table 1.9 Characteristic of IEEE 802.16, IEEE 802.16a, and IEEE 802.16e

Characteristics	IEEE 802.16	IEEE 802.16a	IEEE80.2.16e
Frequency spectrum	10–66GHz	2–11GHz	2–11GHz for fixed 2–6GHz for mobile
Frequency operation	Line of sight	Non line of sight	NLOS
Maximum bit rate	32–134Mbps	1–75Mbps	1–75Mbps
Modulation type	QPSK, 16 QAM, 64 QAM Single carrier	QPSK, 16 QAM, 64 QAM Single carrier OFDM 256 subcarriers OFDMA 2048 subcarriers	QPSK, 16 QAM, 64 QAM
Duplexing	TDD/FDD	TDD/FDD	TDD/FDD
Channel bandwidth	20, 25, and 28MHz	1.25, 1.75, 3.5, 5,8.75, 10, 14, 15MHz	1.25, 1.75, 3.5, 5,8.75, 10, 14, 15MHz
Transmission method	Single carrier	256 OFDM or 2048 OFDM	256 OFDM single-carrier or scalable OFDM (128, 512, 1024, 2048 subcarrier 1.25, 1.75, 3.5, 5,8.75, 10, 14, 15MHz)
Topology	PMP	PMP and mesh	PMP and mesh

1.12 BLUETOOTH

Ericsson Mobile Communication of Sweden began research in 1994 to investigate the use of RF for communications between mobile phones, wireless devices, and PCs. It concluded that all portable devices can be linked using RFs (Bluetooth). To help establish the standard, Ericsson was joined by IBM, Nokia, and Toshiba to form the Bluetooth Special Interest Group (SIG) in 1998. The group grew in 1999 with the addition of Motorola, Microsoft, Lucent, and 3Com. Stated broadly, Bluetooth is the standard by which devices can be connected wirelessly over a maximum distance of 10 meters. Specific applications of Bluetooth include the following:

- Wirelessly connecting keyboards, mice, and other peripherals to a PC.
- Enabling mobile phones to communicate with laptops and personal digital assistants (PDAs) wirelessly. For example, a laptop computer can access the Internet in any location by using a mobile phone that has modem capability.

1.12.1 Bluetooth Topology

Bluetooth offers two types of topologies: Piconet and Scatternet.

- **Piconet topology:** The basic topology of Bluetooth, Piconet is a collection of up to seven devices that are connected in an ad hoc fashion, as shown in Figure 1.25. In this figure, one device is designated as the master, and all others are slaves. The function of the master is to set up a hopping pattern for the slaves and to assign specific time slots for each slave to transmit its data. The slaves communicate exclusively with the master. The master can be connected to seven active slaves simultaneously or 200 inactive or parked slaves. However, any device in a Piconet topology may become the master. Each active device is assigned a 3-bit number (from 1 to 7) that is called the active member address (AMA). Parked devices are assigned an 8-bit number called the parked member address (PMA).

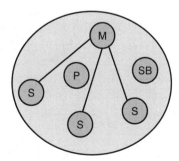

Figure 1.25 Piconet topology

- **Scatternet topology:** Scatternet is a network of Piconets. This allows communication between devices in different Piconets. Figure 1.26 shows a diagram of a Scatternet topology. Two Piconets can create a Scatternet if one slave located in both Piconets acts as a bridge.

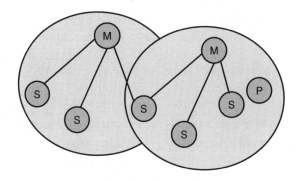

Figure 1.26   Scatternet topology

1.12.2 Bluetooth Protocol Architecture

Figure 1.27 shows the Bluetooth protocol architecture, which is made of numerous layers and protocols that do not follow the TCP/IP protocol or the OSI model.

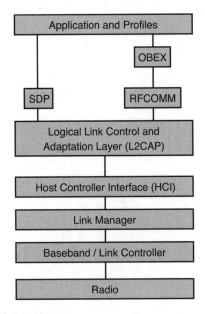

Figure 1.27   Bluetooth protocol architecture

1.12.3 Radio Layer

Bluetooth operates in the 2.40 to 2.480GHz range of the ISM band. The ISM frequency spectrum is divided into 79 channels of 1MHz each. Bluetooth uses FSK for modulation; each signal therefore represents 1 bit, which results in a data rate of 1Mbps. Bluetooth uses FHSS with time-division duplexing (TDD) for communication between slave and master.

1.12.4 Baseband and Data Link Control

The functions of the baseband and data link controller layers are to manage the radio layer operation and physical channels, establish a connection between master and slave, and to determine the state of a station. A station can either be in the standby state or the connection state. In a connection state, the slave and master exchange information, whereas the standby state is the low power-consumption mode.

1.12.5 Link Management Protocol (LMP)

The following list describes the functions of LMP:

- To set up a link between Bluetooth devices
- To negotiate with other devices for selecting packet size for data transmission
- To manage power operations
- Authentication
- Encryption
- To switch the role of slave to master or master to slave
- To close a connection between master and slave

1.12.6 Host Controller Interface (HCI)

The HCI provides an interface between the higher layer of Bluetooth and the lower layer.

1.12.7 Logical Link Control and Adaptation Protocol (L2CAP)

The following list describes the functions of L2CAP:

- To create a logical link between devices.
- To terminate a logical link.
- Negotiating quality of service between Bluetooth devices.
- Multiplexing of multiple channels using a single Bluetooth link. This means multiple protocols such as TCP/IP, OXEB, and SDP can be used simultaneously.
- Segmentation of data packets that exceed the maximum transfer unit of the Bluetooth payload (2755 bytes).
- Reassembly of incoming packets in appropriate order.

1.12.8 RFCOMM

The RFCOMM protocol emulates the RS-232 serial protocol, which is used for transmitting characters to and from applications that are using the serial port for communication.

1.12.9 Service Discovery Protocol (SDP)

The SDP is used by a Bluetooth client to search for a service or for a Bluetooth server to respond to a SDP request.

1.12.10 Object Exchange Protocol (OBEX)

This protocol provides simple file transfers between mobile devices such as sending business cards, personnel calendar entries, and scheduling information

1.12.11 Physical Links

The Bluetooth device support two types of connections: synchronous connection oriented (SCO) and asynchronous connectionless (ACL).

SUMMARY

- The International Standard Organization (ISO) developed the Open Systems Interconnection (OSI) reference model.
- The OSI model consists of seven layers, which, listed from top to bottom, are the application layer, presentation layer, session layer, transport layer, network layer, data link layer, and physical layer.
- The application layer enables the user to access the network applications.
- The presentation layer is responsible for representation of information such as ASCII, encryption, and decryption.
- The function of the session layer is to establish a session between source and destination applications, and to disconnect a session between two applications.
- The function of the transport layer is to ensure that data gets to the destination, to perform error and flow control, and to ensure quality of service.
- The function of the network layer is to deliver information from the source to the destination and route the information.
- The function of the data link layer is to perform framing, error detection, and retransmission.
- The function of the physical layer is to provide an electrical interface, to identify the type of signal, to convert bits to signals (electrical or optical or wireless), and vice versa.
- The IEEE 802 committee developed the standard for the physical and data link layers.

- The IEEE 802 divides the data link layer into two sublayers: Logical Link Control (LLC) and Media Access Control (MAC).

- The IEEE 802.2 is the standard for LLC.

- The IEEE 802.11, IEEE 802.11a, IEEE 802.11b, and IEEE 802.11n are the standards for wireless communications.

- The IEEE 802.16 is the standard for Wireless MAN.

- WLANs use radio frequency (RF) and infrared (IR) signals for transmitting information.

- IEEE 802.11 defines frequency-hopping spread spectrum (FHSS) and direct-sequence spread spectrum (DSSS) for the physical layer using RF signals.

- IEEE 802.11 defines carrier-sense multiple access with collision avoidance (CSMA/CA) and point of coordination function (PCF) for access methods.

- WLANs use ISM or U-NII bands.

- WLAN topologies are managed or unmanaged.

- Managed wireless network topologies are basic service set (BSS) and extended service set (ESS).

- Narrowband signal refers to a signal with a narrow spectrum.

- Spread spectrum signal refers to a signal with range of frequencies.

- IEEE 802.11, IEEE 802.11b, and IEEE 802.11g operate in the ISM band.

- The data rates for IEEE 802.11b are 1, 2, 5.5, and 11Mbps.

- IEEE 802.11b offers three nonoverlapping channels: 1, 6, and 11.

- IEEE 802.11b uses complementary code keying for transmission of information.

- IEEE 802.11g operates in 2.4GHz and uses DSSS and OFDM for transmission of information.

- Wireless MAN is also known as Broadband Wireless Access (BWA).

- Wireless MAN defines three types of stations: base station (BS), subscriber station (SS), and repeater.

- IEEE 802.16 supports point-to-multipoint (PMP) and mesh topologies.

- Transmission from subscriber to the BS is called an uplink transmission.

- The IEEE defines two types of physical layers: IEEE 802.16 and IEEE 802.16a.

- The MAC layer is divided into the three sublayers: the convergence sublayer, the common part sublayer, and the security sublayer:

  - Convergence sublayer (CS): IEEE 802.16 is designed to carry various types of protocol data units (PDUs) such as ATM, IPv4, IPv6, and Ethernet.

  - MAC common part sublayer (CPS): The IEEE 802.16 MAC layer uses a connection-oriented PMP transmission.

- Security sublayer: The security sublayer performs authentication between the BS and each SS, encryption, and decryption.
- Each SS has a 48-bit MAC address and a 16-bit connection identifier (CID).
- Basic connection: The base station and subscriber stations use basic connections to exchange short and urgent management messages such as SS basic capability requests and responses.
- Bluetooth uses Piconet and Scatternet topologies.
- Master devices control the operation of Piconet.
- Masters set the hopping pattern.
- In a Piconet, a station can be a master station, slave station, parked station, or standby station.
- Bluetooth operates in the ISM frequency spectrum.
- Bluetooth uses FHSS with time-division multiplexing (TDM) for communication between master and slave.
- Bluetooth supports synchronous connection oriented (SCO) and asynchronous connectionless links (ACL).
- Bluetooth supports 1/3FEC, 2/3FEC, and ARQ for error correction.
- Bluetooth devices use three different addresses: active member address, parked member address, hardware address.

REFERENCES

1. Elahi and Elahi, *Data, Network, & Communications Technology* (Thomson, 2006)
2. Bluetooth Specifications, Bluetooth SIG, at www.bluetooth.com/

CHAPTER 2

ZIGBEE WIRELESS SENSOR AND CONTROL NETWORK

INTRODUCTION

ZigBee is a new standard developed by the ZigBee Alliance for personal-area networks (PANs). Consisting of more than 270 companies (including Freescale, Ember, Mitsubishi, Philips, Honeywell, and Texas Instruments), the ZigBee Alliance is a consortium that promotes the ZigBee standard for a low-rate/low-power wireless sensor and control network. The ZigBee protocol stack is built on top of IEEE 802.15.4, which defines the Media Access Control (MAC) and physical layers for low-rate wireless personal-area network (LR-WPAN). The ZigBee standard offers a stack profile that defines the network, security, and application layers. Developers are responsible for creating their own application profiles or integrating with the public profiles that were developed by the ZigBee Alliance. The ZigBee specification is an open standard that allows manufacturers to develop their own specific applications that require low cost and low power.

The ZigBee specification has undergone multiple modifications. The major milestones in its revision history are listed here:

- In 2004, the ZigBee Alliance published its first specification, which supported a home control lighting profile. However, the ZigBee Alliance no longer supports the 2004 specification.

- In February 2006, the ZigBee Alliance published the ZigBee Stack 2006, which contained modifications to ZigBee 2004.

- In October 2007, the ZigBee Alliance published two feature sets called ZigBee and ZigBee PRO. The ZigBee feature set is interoperable with ZigBee PRO. If a network is based on the ZigBee PRO stack, devices from the ZigBee feature set stack can join the network as end devices. Likewise, if a network is based on the ZigBee stack, ZigBee PRO devices can join the network as end devices.

The ZigBee feature set is backward compatible with ZigBee 2006; a ZigBee feature set device can join in a ZigBee 2006 network and vice versa.

The ZigBee Alliance developed the following application profiles:

- **Smart energy:** ZigBee can be used to quickly read electrical, gas, and water meters. The ZigBee smart energy network enables wireless communication between the advanced metering infrastructure (AMI) and the home-area network; that is, the smart energy network will connect home appliances with the utility company for improving energy efficiency and managing peak demand.

- **Commercial building automation:** In a commercial building, ZigBee can be an integral tool in building maintenance. ZigBee wireless can be used to monitor smoke-detector operation and fire-door position. Suppose that a high-rise building contains 50 floors, with each floor having 50 rooms, and each room is equipped with a smoke detector. For safety reasons, each smoke detector must be tested every month. This requires checking 2,500 rooms! Instead of requiring that someone manually test the 2,500 smoke detectors, ZigBee allows a central station to remotely monitor each smoke detector. A ZigBee device may also be used to turn on and off a light without using any wire.

- **Home automation:** ZigBee home automation profile defines devices that are used for residential and commercial applications. ZigBee can be used to remotely control lighting, heating, cooling, and door-locking mechanisms. It can also remotely monitor smoke detectors and home security systems

- **Personal, home, and hospital care (PHHC):** This profile is used for monitoring the personal health of a patient at home without limiting a patient's mobility. For example, it can remotely monitor blood pressure and heart rate.

- **Telecom applications:** Embedding a ZigBee device into a mobile phone or PDA creates a new device called a ZigBee mobile device. A ZigBee mobile device can be used to communicate with other ZigBee devices. Users of ZigBee mobile devices can send and receive messages and share ring tones, contacts and images. More important, the ZigBee mobile device can even communicate with ZigBee devices that use different application profiles. For example, a ZigBee mobile device can be used to alert emergency services when a PHHC-enabled device detects a critical problem with the patient's health.

- **Remote control for consumer electronics (ZigBee RF4CE):** Currently, most remote controllers are using infrared (IR) technology, which requires line of sight; ZigBee RF4CE is a protocol that uses radio frequency (RF) to replace IR technology for remote controllers used in consumer electronics.

- **Industrial process monitoring and control:** ZigBee offers solutions for wireless sensor and control. Therefore, it can be used for monitoring and controlling industrial processes without using wire. For example, in inventory tracking, each piece of equipment can be tagged with a wireless sensor and can then be located by a ZigBee node. This process is called radio frequency identification (RFID). ZigBee can monitor machine condition and the performance of operating equipment within a plant; ZigBee can record and transmit such critical information as temperature, pressure, flow, tank level, humidity, and vibration.

2.1 ZIGBEE NETWORK CHARACTERISTICS

Several standards currently exist for wireless networks, including Bluetooth, WiFi, and WiMax. ZigBee is a new standard for wireless sensor and control networks. It has the following characteristics:

- Low battery consumption. A ZigBee end device should operate for months or even years without needing its battery replaced.
- Low cost.
- Low data rate. The maximum data rate for a ZigBee device is 250Kbps.
- Easy to implement.
- Supports up to 65,000 nodes connected in a network.
- ZigBee can automatically establish its network.
- ZigBee uses small packets compared with WiFi and Bluetooth.

Table 2.1 shows a comparison of ZigBee characteristics with those of WiFi and Bluetooth.

Table 2.1   ZigBee, Bluetooth, and WiFi Characteristics

	WiFi IEEE 802.11	Bluetooth IEEE 802.15.1	ZigBee IEEE 802.15.4
Application	Wireless LAN	Cable replacement	Control and monitor
Frequency bands	2.4GHz	2.4GHz	2.4GHz, 868MHz, 915MHz
Battery life (days)	0.1–5	1–7	100–7,000
Nodes per network	30	7	65,000
Bandwidth	2–100Mbps	1Mbps	20–250Kbps
Range (meters)	1–100	1–10	1–75 and more
Topology	Tree	Tree	Star, tree, cluster tree, and mesh
Standby current	$20 * 10{-}3$ amps	$200 * 10{-}6$ amps	$3 * 10{-}6$ amps
Memory	100KB	100KB	32–60KB

Table 2.1 shows that Bluetooth, while similar in functionality to ZigBee, does not offer the range of topologies, and its standby current is nearly 70 times more than ZigBee. Of the three wireless networks under comparison, ZigBee is the only one that offers mesh topology. In addition, a ZigBee end device can be in sleep mode and still keep its association with its network. ZigBee is considered a more sophisticated network when compared to either Bluetooth or WiFi.

2.2 ZIGBEE DEVICE TYPES

A ZigBee network consists of ZigBee nodes (devices). The node architecture is shown in Figure 2.1. A node consists of a microcontroller, a transceiver, and an antenna. A ZigBee node uses stack profiles, which are developed by software. A node can be used for a wide variety of applications—for example, lighting control, smoke-detector, and home-security monitoring. Therefore, a node can support multiple subunits, and each subunit has an application object that describes the subunit function. A node can operate as either a full-function device (FFD) or reduced-function device (RFD). An FFD can perform all the tasks that are defined by the ZigBee standard, and it operates in the full set of the IEEE 802.15.4 MAC layer. An RFD performs only a limited number of tasks.

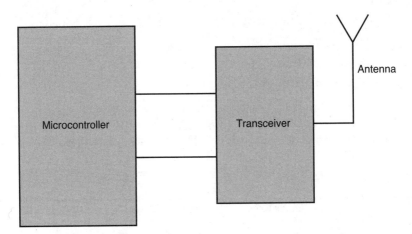

Figure 2.1　Architecture of ZigBee node

- **Coordinator:** A coordinator is an FFD and responsible for overall network management. Each network has exactly one coordinator. The coordinator performs the following functions:

  Selects the channel to be used by the network

  Starts the network

Assigns how addresses are allocated to nodes or routers

Permits other devices to join or leave the network

Holds a list of neighbors and routers

Transfers application packets

- **End device:** An end device can be an RFD. An RFD operates within a limited set of the IEEE 802.15.4 MAC layer, enabling it to consume less power. The end device (child) can be connected to a router or coordinator (parent). It also operates at low duty-cycle power, meaning it consumes power only while transmitting information. Therefore, ZigBee architecture is designed so that an end device transmission time is short. The end device performs the following functions:

Joins or leaves a network

Transfers application packets

- **Router:** A router is an FFD. A router is used in tree and mesh topologies to expand network coverage. The function of a router is to find the best route to the destination over which to transfer a message. A router performs all functions similar to a coordinator except the establishing of a network.
- **ZigBee trust center (ZTC):** The ZigBee trust center is a device that provides security management, security key distribution, and device authentication.
- **ZigBee gateway:** The ZigBee gateway is used to connect the ZigBee network to another network, such as a LAN, by performing protocol conversion.

2.3 ZIGBEE TOPOLOGIES

ZigBee uses the IEEE 802.15.4 2003 specification for its physical layer and MAC layer. IEEE 802.15.4 offers star, tree, cluster tree, and mesh topologies; however, ZigBee supports only star, tree, and mesh topologies.

It uses an association hierarchy; a device joining the network can either be a router or an end device, and routers can accept more devices.

- **Star topology:** The star topology consists of a coordinator and several end devices (nodes), as shown in Figure 2.2. In this topology, the end device communicates only with the coordinator. Any packet exchange between end devices must go through the coordinator. The disadvantage of this topology is the operation of the network depends on the coordinator of the network, and because all packets between devices must go through coordinator, the coordinator may become bottlenecked. Also, there is no alternative path from the source to the destination. The advantage of star topology is that it is simple and packets go through at most two hops to reach their destination.

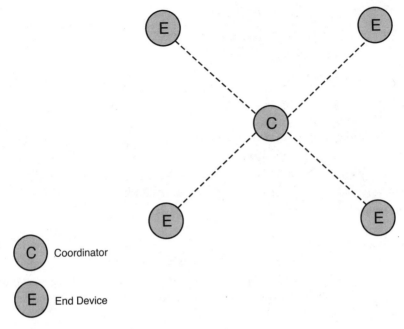

Figure 2.2 Star topology

- **Tree topology:** In this topology, the network consists of a central node (root tree), which is a coordinator, several routers, and end devices, as shown in Figure 2.3. The function of the router is to extend the network coverage. The end nodes that are connected to the coordinator or the routers are called children. Only routers and the coordinator can have children. Each end device is only able to communicate with its parent (router or coordinator). The coordinator and routers can have children and, therefore, are the only devices that can be parents. An end device cannot have children and, therefore, may not be a parent. A special case of tree topology is called a cluster tree topology.

The disadvantages of tree topology are

  a. If one of the parents becomes disabled, the children of the disable parent cannot communicate with other devices in the network.

  b. Even if two nodes are geographically close to each other, they cannot communicate directly.

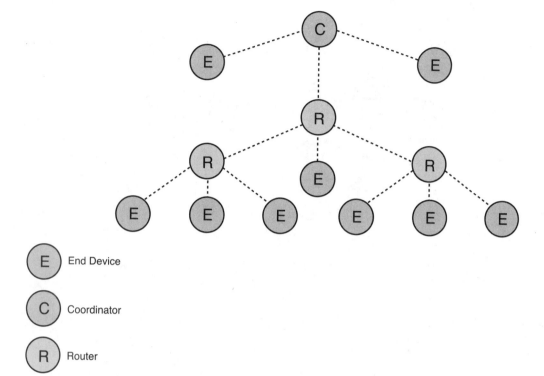

Figure 2.3   Tree topology

- **Cluster tree topology:** A cluster tree topology is a special case of tree topology in which a parent with its children is called a cluster, as shown in Figure 2.4. Each cluster is identified by a cluster ID. ZigBee does not support cluster tree topology, but IEEE 802.15.4 does support it.

- **Mesh topology:** Mesh topology, also referred to as a peer-to-peer network, consists of one coordinator, several routers, and end devices, as shown in Figure 2.5. The following are the characteristics of a mesh topology:

  A mesh topology is a multihop network; packets pass through multiple hops to reach their destination.

  The range of a network can be increased by adding more devices to the network.

  It can eliminate dead zones.

  A mesh topology is self-healing, meaning during transmission, if a path fails, the node will find an alternate path to the destination.

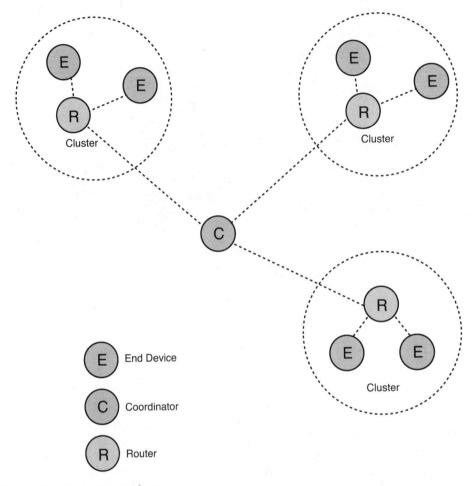

Figure 2.4 Cluster tree topology

Devices can be close to each other so that they use less power.

Adding or removing a device is easy.

Any source device can communicate with any destination device in the network.

Compared with star topology, mesh topology requires greater overhead.

Mesh routing uses a more complex routing protocol than a star topology.

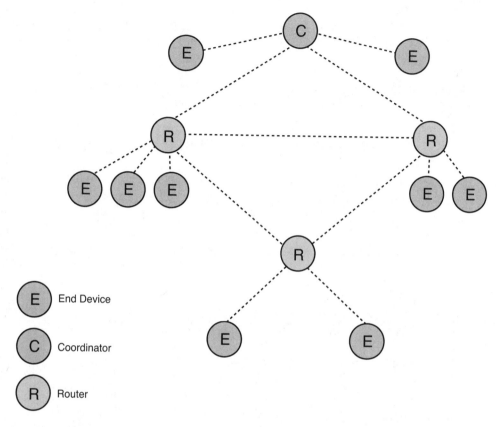

Figure 2.5   Mesh topology

2.4 END DEVICE (NODE) ADDRESSING

When a device joins a ZigBee network, the ZigBee coordinator or router assigns a logical 16-bit address to the device. Also, each device has a 64-bit IEEE address, and no two devices can have the same IEEE address in the entire world. Short addresses (16 bits) can be used by devices in a network. The advantage of using the 16-bit address is to extend the life of the battery. A 16-bit address reduces the size of the frame. A smaller frame size results in shorter transmission time. Less transmission time means greater battery life. The disadvantage of using the 16-bit address is that two nodes in different networks can have the same address.

2.5 DEPTH OF A NETWORK, NUMBER OF CHILDREN, AND NETWORK ADDRESS ALLOCATION

- **Depth of a network:** The depth of a network is determined by the number of routers (hops) from the coordinator to the farthest device, where farthest is defined by number of hops. In a star topology, the depth of a network is one.

- **Number of children:** The number of end devices (children) that are connected to a router or coordinator. The coordinator sets the maximum number of children connected to a router.

- **Address allocation:** In a tree topology, each coordinator holds information about the network, such as the maximum number of children (the number of end devices connected to each router), maximum number of routers, and uses this information to assign an address to each router. The routers, then, assign the addresses to their respective end devices (children). In a mesh topology, however, each router assigns a random address to its respective end devices.

2.6 ZIGBEE PROTOCOL ARCHITECTURE

Figure 2.6 shows the ZigBee protocol architecture. The ZigBee Alliance developed the Zig-Bee device object (ZDO), the application support sublayer (APS), the network layer, and security management. IEEE 802.15.4 is used for the MAC layer and physical layer.

The ZigBee protocol architecture is divided into three sections, as follows:

- IEEE 802.15.4, which consists of the MAC and physical layers.
- ZigBee layers, which consist of the network layer, the ZigBee device object (ZDO), the application sublayer, and security management.
- Manufacturer application: Manufacturers of ZigBee devices can use the ZigBee application profile or develop their own application profile.

2.6.1 Physical Layer

The physical layer performs modulation on outgoing signals and demodulation on incoming signals. It transmits information and receives information from a source. Table 2.2 shows the physical layer frequency band, data rate, and channel numbers.

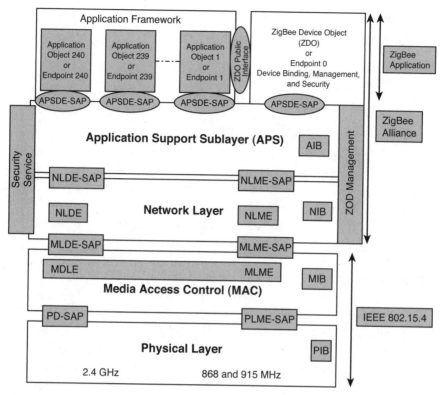

Figure 2.6   ZigBee protocol architecture

Table 2.2   Physical Layer Frequency Band

Frequency Band	Country	Data Rate	Channel Numbers
868.3MHz	European countries	20Kbps	0
902–928 MHz	United States	40Kbps	1–10
2.405GHz	Worldwide	250Kbps	11–26

2.6.2 Media Access Control (MAC) Layer

The functions of the MAC layer are to access the network by using carrier-sense multiple access with collision avoidance (CSMA/CA), to transmit beacon frames for synchronization, and to provide reliable transmission.

2.6.3 Network Layer

The network layer is located between the MAC layer and application support sublayer. It provides the following functions:

- Starting a network
- Managing end devices joining or leaving a network
- Route discovery
- Neighbor discovery

2.6.4 Application Support Sublayer (APS)

The application support sublayer (APS) provides the services necessary for application objects (endpoints) and the ZigBee device object (ZDO) to interface with the network layer for data and management services. Some of the services provided by the APS to the application objects for data transfer are request, confirm, and response. Furthermore, the APS provides communication for applications by defining a unified communication structure (for example, a profile, cluster, or endpoint).

- **Application object (endpoint):** An application object defines input and output to the APS. For example, a switch that controls a light is the input from the application object, and the output is the light bulb condition. Each node can have 240 separate application objects. An application object may also be referred to as an endpoint (EP). Figure 2.7 shows an example of home control lighting.
- **ZigBee device object (ZDO):** A ZigBee device object performs control and management of application objects. The ZDO performs the overall device management tasks:

   Determines the type of device in a network (for example, end device, router, or coordinator)

   Initializes the APS, network layer, and security service provider

   Performs device and service discovery

   Initializes coordinator for establishing a network

   Security management

   Network management

   Binding management

- **End node:** Each end node or end device can have multiple EPs. Each EP contains an application profile, such as home automation, and can be used to control multiple devices or a single device. More to the point, each EP defines the communication functions within a device. As shown in Figure 2.7, the bedroom switch controls the bedroom light, and the remote control is used to control three lights: bedroom, hallway1, and hallway2.

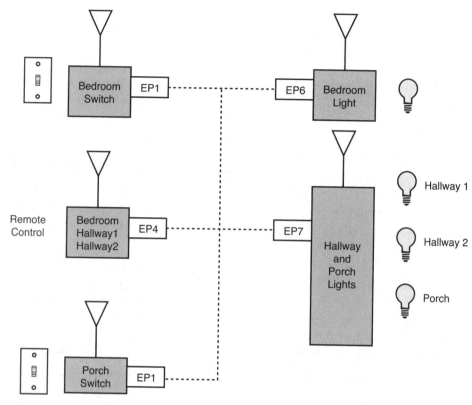

Figure 2.7   Home control lighting

- **ZigBee addressing mode:** ZigBee uses direct, group, and broadcast addressing for transmission of information. In direct addressing, two devices communicate directly with each other. This requires that the source device has both the address and endpoint of the destination device. Group addressing requires that the application assign a group membership to one or more devices. A packet is then transmitted to the group address in which the destination device lies. The broadcast address is used to send a packet to all devices in the network.

2.7 ZIGBEE AND ZIGBEE PRO FEATURE SETS

In October 2007, the ZigBee Alliance published two feature set specifications for ZigBee: ZigBee and ZigBee PRO. The ZigBee feature set is nearly the same as the ZigBee 2006 specification; however, it does offer a few new features. ZigBee PRO offers several significant improvements such as security and the capability to self-form and self-heal the network. It is targeted for use in building automation and environmental and industrial applications that contain more than 30 nodes. ZigBee PRO is based on a mesh topology and is a beaconless network. The following are its characteristics:

- **Addressing:** ZigBee PRO uses a mesh topology. Any device joining the network requires a network address. ZigBee PRO uses a stochastic addressing method, which means a ZigBee device randomly picks up an address when joining the network. Also, the device announces its address on the network by using the device-annc command. The ZigBee network layer provides address conflict resolution based on the MAC address if two nodes have the same network addresses. Stochastic addressing eliminates the need for the parent to maintain an address table for assignment to children.

- **Link management:** In a mesh topology, each node can communicate with its neighbor. The node has the ability to evaluate the quality of its neighbor's links and to select the best one for transmission of the packet.

- **Frequency agility:** ZigBee PRO selects the best available channel during startup of a network but, during network operation, if any node detects interference due to frequency conflict or noise, it will report the occurrence to the channel manager. (A channel manager can be a dedicated device or trust center.) When the channel manger receives the reports from several nodes on the network, it selects another channel for network operation and informs the nodes of the switch.

- **Group addressing:** ZigBee PRO provides group addressing; a single packet can reach a group of devices.

- **Commissioning:** It is a tool that is used by the ZigBee device installer to install ZigBee devices.

- **Compatibility:** The ZigBee PRO stack identifier is two and is advertised in its beacon frame. Any end device can join ZigBee PRO if it uses the standard security mode.

- **Asymmetric link:** The link between two nodes is usually asymmetric, meaning the quality of the link in both directions is not the same. The quality of the link is represented by the link cost. Figure 2.8 shows a ZigBee network. The link cost from node A to B is different from B to A (asymmetric link) because A and B do not have the same transmission powers and receiver sensitivities. When the nodes A and E use the same route B–D to exchange information, it is considered symmetric. In asymmetric routing, the destination node uses a different path to transmit to the source. In Figure 2.8, the source A uses the B–D path to send a packet to E, and E use the path C to send the packet to A.

- **Fragmentation:** ZigBee PRO provides fragmentation of large packets into smaller packets for transmission. The destination node will reassemble the packets.

- **Power management:** In ZigBee PRO, only the end devices are powered by batteries; the routers and coordinator use main power. ZigBee PRO allows end devices to go into sleep mode so that they consume less power. While the end device is in sleep mode, the node will miss any network key updates from the trust center. When the device wakes up, it uses its link key to send a message to the trust center to obtain an updated network key.

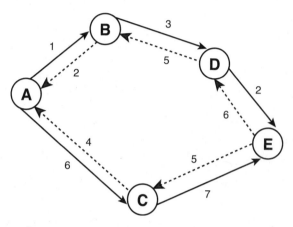

Figure 2.8 Asymmetric link

- **Routing:** ZigBee PRO offers two types of routing: many-to-one routing and multicast routing. Many-to-one routing is used for the network with a concentrator.
- **Security:** ZigBee PRO offers two security modes, as follows:
  - *Standard mode:* In standard mode, the device is permitted to use the network key and link key. The network key is a standard key and all devices share the same key. In this mode, devices do not require authentication to join a network. The trust center, master key, and SKKE are optional. The trust center for ZigBee PRO standard security mode is used for transporting the network key.
  - *High-security mode:* In this mode, three keys are permitted for use: the network, link, and master keys. The trust center and SKKE are mandatory. The trust center uses the transport key command to transport the link and network keys to the devices in the network. The device is required to perform authentication with its parent, and it is required to perform authentication between neighbors.
- **Trust center:** ZigBee PRO should have a trust center; the trust center can be a router, coordinator, concentrator, or specific device.

Table 2.3 shows a comparison of ZigBee 2006, ZigBee, and ZigBee PRO feature sets.

Table 2.3   Comparison of 2006, ZigBee, and ZigBee PRO Feature Sets

ZigBee Features	2006 Specification	ZigBee Feature Set Specification	ZigBee PRO Feature Set Specification
Network coordinator selects best channel at startup.	Yes	Yes	Yes
During operation can detect interference and change channel operation.	No	Yes	Yes
Distributed address assignment.	Yes	Yes	No
Stochastic address assignment.	No	No	Yes
Supports group addressing.	Yes	Yes	Yes
Many-to-one routing.	No	No	Yes
128-bit Advanced Encryption Standard (AES) with message integrity code (MIC).	Yes	Yes	Yes
Trust center can be any device or coordinator in the network.	Coordinator	Coordinator	Any device
Network scale up limited to address assignment scheme.	Yes	Yes	No
Message fragmentation is permitted.	No	Yes	Yes
Supports buffering for message fragmentation.	No	Yes	Yes
Supports commissioning tool.	Yes	Yes	Yes
Device keeps information about its neighbor devices.	No	No	Yes
Offers high-security mode.	No	No	Yes
Network topologies.	Tree and mesh	Tree and mesh	Mesh

SUMMARY

This chapter presented an overview of ZigBee, its applications, and its characteristics. The following are the key concepts that were described in the chapter:

- ZigBee, a wireless sensor and control network, was developed by the ZigBee Alliance.
- ZigBee applications can be used in home automation, commercial building automation, personal home health care, smart energy, and industrial process monitoring.
- The first ZigBee specification was published in 2004 and supported home control lighting; the ZigBee Alliance no longer supports the 2004 specification.
- In 2006, the ZigBee Alliance published the ZigBee 2006 specification, which was a modification of ZigBee 2004 specification.
- In 2007, ZigBee published ZigBee and ZigBee PRO feature sets.
- ZigBee defines three main types of devices: the coordinator, router, and end device. In addition, devices can act as a trust center or gateway.

- The coordinator is a full-function device (FFD). It performs the critical function of controlling the network, starting a network, and permitting other devices to join or leave the network.
- The trust center performs authentication of devices joining the network, security management, and key distribution.
- ZigBee offers star, tree, and mesh topologies.
- It uses layer architecture for its protocol.
- It uses IEEE 802.15.4 for its physical and MAC layers.
- ZigBee protocol architecture consists of the application, application support sublayer, and network layers
- It offers direct, group, and broadcast addressing.
- A ZigBee end device can have 240 endpoints, where each endpoint may represent different application.
- ZigBee PRO uses a mesh topology, and it is a beaconless network.
- ZigBee PRO offers stochastic addressing, group addressing, asymmetric links, fragmentation, and frequency agility.
- ZigBee PRO offers many-to-one routing and multicast routing.
- ZigBee PRO offers standard and high-security modes.
- ZigBee PRO security level is set to 5.
- ZigBee PRO provides link management.
- ZigBee PRO end devices are only powered by battery.
- The ZigBee 2007 specification supports tree and mesh topologies.
- ZigBee PRO selects the best channel at startup and during operation.

REFERENCES

1. ZigBee Alliance, ZigBee Specification Document 053474r17, 2008
2. Daintree Network, "Comparing ZigBee Specification Versions," www.daintree.net/resources/spec-matrix.php
3. "How Does ZigBee Compare with Other Wireless Standards?" www.stg.com/wireless/ZigBee-comp.html
4. Craig, William C. " ZigBee: Wireless Control That Simply Works," ZigBee Alliance, 2003
5. Ember Corporation, EmberZNet Application Developer's Reference Manual, 2008
6. IEEE Std 802.15.4 2003
7. IEEE Std 802.15.4 2006

CHAPTER 3

IEEE 802.15.4 PHYSICAL LAYER

INTRODUCTION

IEEE 802.15.4 developed the physical and Media Access Control (MAC) layers for low-rate personal-area network (LR-PAN). The first specification was published in 2003 and was later revised in 2006. However, the revised 2006 specification is backward compatible with the 2003 specification.

The physical layer operates in the Industrial, Scientific, and Medical (ISM) band and, in Europe, in the 868 to 868.6MHz band known simply as the 868MHz band. It provides data transmission and reception, modulation of outgoing signals, and demodulation of incoming signals, and it manages the function of the transceiver (transmitter and receiver). The physical layer also performs the following functions:

- **Activation and deactivation of the transceiver.**
- **Clear channel assessment (CCA):** Checks whether the channel is clear.
- **Energy detection (ED):** Measures energy level of the channel.
- **Link quality indication (LQI):** Indicates the quality of incoming packets.
- **Channel selection:** As channels are divided into pages, IEEE 802.15.4 offers 27 channels on page 0 and 11 channels each on pages 1 and 2.

3.1 FREQUENCY BAND, DATA RATE, AND CHANNEL NUMBERING

The frequency band and data rate depend on a country's regulations and standards. In the European countries, the European Standard Institute recommends the frequency band of 868 to 868.6MHz. Whereas in the countries of North America, Australia, and New Zealand, the FCC recommends the 915MHz and 2.4GHz bands. As shown in Table 3.1, the 915MHz or I-band ranges from 902MHz to 928MHz, and the 2.4GHz band, or S-band, operates from 2.4GHz to 2.48GHz.

The physical layer uses direct-sequence spread spectrum (DSSS) for transmission of information. This is a process by which the physical layer converts information symbols to 32-bit chip bits, where each chip bit is modulated for transmission. In its 2006 revision, IEEE 802.15.4 added physical layer support for optional modulation techniques on the various frequency bands. Table 3.2 shows the IEEE 802.15.4 frequency bands and data rates; the optional frequencies and corresponding data rates are supported by the IEEE802.15.4 2006 specification.

Table 3.1 IEEE 802.15.4 Frequency Band and Modulation Types

Frequency Band	Frequency Range	Modulation	Data Rate	Symbol Rate
868MHz	868–868.6MHz	BPSK	20Kbps	20K/s
868MHz optional	868–868.6MHz	O-QPSK	100Kbps	25K/s
868MHz optional	868–868.6MHz	ASK	250Kbps	12.5K/s
915MHZ	902–928MHz	BPSK	40Kbps	40K/s
915MHz optional	902–928MHz	O-QPSK	250Kbps	62.5K/s
915MHz optional	902–928MHz	ASK	250Kbps	50 K/s
2.4GHz	2400–2483.5MHz	O-QPSK	250Kbps	62.5K/s

- **Channel numbering:** IEEE 802.15.4 uses 32 bits to represent the channel number. This 32-bit number is divided into two segments, as shown here:

$$B31B30\ B29\ B28B27\ B26B25\ldots\ldots\ldots B7B6\ B5B4\ B3B2\ B1B0$$

$\longleftarrow$ *Page Number* $\longrightarrow$ $\longleftarrow$ *Channel Number* $\longrightarrow$

The leftmost 5 bits—$B_{31}B_{30}B_{29}B_{28}B_{27}$—represent the page number. This allows for the designation of 32 pages. Each bit of $B_{26}B_{25}\ldots B_7B_6\ B_5B_4\ B_3B_2\ B_1B_0$ represents a channel number within the page. Because the physical layer of a device can support multiple channels within a page, each channel must be represented by a distinct bit. Table 3.3 details the relationship between page number, frequency band, modulation, and channel number.

Table 3.2   IEEE 802.15.4 Frequency Bands, Modulations, and Channel Numbers

Page Number	Frequency Band	Data Rate	Modulation	Channel Number
0	868MHz	20Kbps	BPSK	0
0	915MHz	40Kbps	BPSK	1–10
0	2.4GHz	250Kbps	O-QPSK	11–26
1	868MHz	250Kbps	ASK	1
1	915MHz	250Kbps	ASK	1–10
2	868MHz	100Kbps	O-QPSK	0
2	915MHZ	250Kbps	O-QPSK	1–10
3–31	Reserved			Reserved

As shown in Table 3.2, page 0 offers 27 channels, whereas pages 1 and 2 each offer 11 channels. To increase the number of channels, the physical layer offers different types of modulations in the same frequency band. Table 3.2 details the modulation options available to the various page number/frequency band combinations. As is shown, the 868MHz and 915MHz bands may use binary phase-shift keying (BPSK), amplitude-shift keying (ASK), or offset-quadrature phase-shift keying (O-QPSK) for modulation; the 2.4GHz band uses only O-QPSK for modulation. The advantage of the 868MHz and 915MHz bands is that they have a longer range and use less power than their 2.4GHz counterpart. However, the 2.4GHz band has a higher data rate.

Although page zero offers a total of 27 channels, the 868MHz band offers only 1 channel (channel 0). Of the remaining channels, the 915MHz band offers 10 (numbered 1 through 10), and the 2.4GHz band offers 16 (numbered 11 to 26). This gives a total of 27 channels over which ZigBee can operate. Table 3.3 shows the frequency bands with their corresponding channel numbers and center frequency.

Table 3.3   Frequency Band with Channel Numbers and Center Frequency 868MHz Band

2.4GHz Band

Channel Number	Center Frequency	Channel Number	Center Frequency
0	868.3MHz	11	2405MHz
		12	2410MHz
915MHz Band		13	2415MHz
		14	2420MHz
Channel Number	**Center Frequency**	15	2425MHz
		16	2430MHz
1	906MHz	17	2435MHz
2	908MHz	18	2440MHz
3	910MHz	19	2445MHz
4	912MHz	20	2450MHz
5	914MHz	21	2455MHz
6	916MHz	22	2460MHz
7	918MHz	23	2465MHz
8	920MHz	24	2470MHz
9	922MHz	25	2475MHz
10	924MHz	26	2480MHz

- **IEEE 802.15.4 and IEEE802.11 channels:** IEEE 802.11 b/g operates in the ISM band and has three nonoverlapping channels: 1, 6, and 11. Figure 3.1 shows 802.11 b/g channels and ZigBee channels. Using the information in Figure 3.1, a user can select the proper channels for WLANs and wireless sensors. As shown in Figure 3.1, the best channels over which ZigBee may operate are 15, 20, 25, and 26 because they are the least likely to have interference from the 802.11 channels.

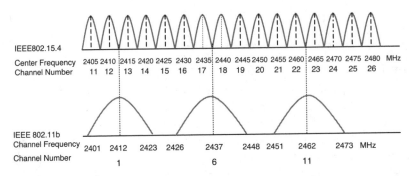

Figure 3.1   IEEE 802.15.4 and IEEE 802.11 channels

3.2 PHYSICAL LAYER SERVICES

The physical layer provides services to the MAC layer and the transceiver through a point of interface called the service access point (SAP). The two types of services provided by the physical layer are data services and management services. The physical layer provides data services through the physical data service access point (PD-SAP) and management services via the physical layer management entity (PLME). The physical layer also contains a database, known as the Physical Layer information base (PIB), which holds the physical layer attributes. The MAC layer management entity (MLME) communicates with the PLME. The exchange of information between layers is called a primitive. Each of these relationships and interfaces can be seen in Figure 3.2.

3.2.1 Physical Layer Data Services

The physical layer uses the following data service primitives to transfer data from the MAC layer to the physical layer and vice versa.

- PD-Data.request
- PD-Data.confirm
- PD-Data.indication

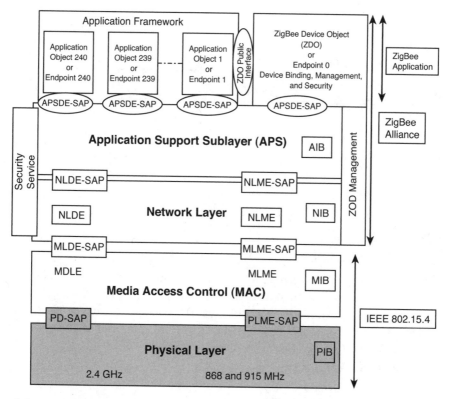

Figure 3.2   Interface between the physical layer and MAC layer

PD-Data.request

A PD-Data.request is generated by the MAC layer and issued to the physical layer. The PD-Data.request transfers the MAC layer information frame known as a protocol data unit (MPDU) to the physical layer. The PD-Data.request primitive can be seen in Figure 3.3. The physical layer adds its header information to the MPDU and transmits the physical protocol data unit (PPDU), as shown in Figure 3.4. When the physical layer completes its transmission, it then transmits the PD-Data.confirm primitive to inform the MAC layer of the transmission status (for example, success or failure). Figure 3.5 shows the PD-Data.confirm.

Figure 3.3   PD-Data.request primitive

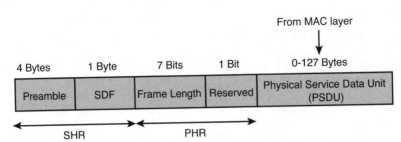

Figure 3.4   Physical layer protocol data unit (PPDU)

Figure 3.5   PD-Data.confirm primitive

- **Preamble:** The preamble is used for synchronization by the receiver. It contains 32 bits for all bands and modulations with the exception of those bands using ASK modulation. When using ASK modulation, the preamble contains 30 bits for the 915MHz band and 40 bits for the 868MHz band. Together, the preamble and SFD comprise the synchronous header (SHR).

- **Start frame delimiter (SFD):** The SFD indicates the end of the preamble and is represented as 111100101.

- **Frame length:** The frame length indicates the length of the frame. Generally, a frame length of 5 indicates that the frame is an acknowledge frame, whereas a length of 9 to its maximum value of 127 indicates that the number of bytes in the payload is data.

- **Physical layer service data unit (PSDU):** The PSDU is the actual data unit passed to the physical layer.

PD-Data.confirm

The PD-Data.confirm command, generated by the physical layer and issued to the MAC layer, indicates the status of the PD-Data.request. The status field contains success, receiver on, or transmitter off. Figure 3.6 shows the PD-Data.request and the PD-Data.confirm process. The MAC layer sends a PD-Data.request containing the MPDU to the physical layer for transmission. The physical layer adds its header and transmits the PPDU to the destination. Finally, when the physical layer has completed the transmission of the PPDU, it sends a PD-Data.confirm to inform the MAC layer of the status of transmission. The confirmation primitive does not indicate that the PPDU packet was received by the destination; it merely indicates that it was sent successfully.

PD-Data.indication

The PD-Data.indication command is sent by the physical layer to the MAC layer. It indicates transfer of the MPDU to the MAC layer from the physical layer. Figure 3.7 shows the PD-Data.indication primitive.

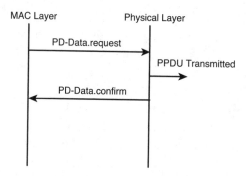

Figure 3.6   PD-Data.request and confirm process

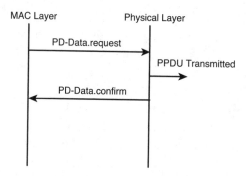

Figure 3.7   PD-Data.indication primitive

- **PSDU length:** This is the number of bytes in the data unit.
- **Link quality indication (LQI):** The LQI is used to indicate the quality of the packets received by the MAC layer. The link quality is derived from a combination of the energy level and the signal-to-noise ratio of the incoming signal.

3.2.2 Physical Layer Management Services

The physical layer management services are provided by the PLME and perform following functions:

- Clear channel assessment (CCA)
- Energy detection (ED)
- Link quality indication (LQI)
- Channel frequency selection
- Enable and disable transceiver

3.2.3 Clear Channel Assessment (CCA)

The CCA is performed by the physical layer to determine whether a channel is clear. The results are sent to the MAC layer for carrier-sense multiple access with collision avoidance (CSMA/CA). For channel assessment, the physical layer uses the PLME-CCA.request primitive:

- **PLME-CCA.request:** The MAC layer requests channel assessment from the physical layer by sending the PLME-CCA.request to the physical layer. This causes the physical layer to perform a channel assessment check. The PLME-CCA.request does not have any parameters.
- **PLME-CCA.confirm:** When the physical layer completes the channel assessment, it sends the results to the MAC layer in the form of the PLME-CCA.confirm to indicate the status of channel (for example, busy, idle, or transmitter off).

3.2.4 Energy Detection (ED)

Energy measurement is used by the network layer of a device to select the best channel available for operation. The full-function device (FFD) scans the selected channels to measure each channel's energy level. This information is used by the coordinator to select the proper channel before indicating to other devices that it is a coordinator. The physical layer uses the PLME-ED.request for scanning the energy levels of specific channels:

- **PLME-ED.request:** This primitive is generated by the MLME of the MAC layer and sent to the PLME of the physical layer to request a channel scan for the purpose of energy-level detection. When the physical layer completes the current channel scan, it sends the result of the scan to the MAC layer by means of the PLME-ED.confirm. The PLME-ED.confirm contains the status of the request, (for example, success or transmitter off) and the power level of the radio frequency for the channel.
- **PLME-SET-TRX-STATE.request:** This command is generated by the MAC layer and issued to the physical layer to change the state of transceiver. The states include transceiver disabled, transmitter on, and receiver on. When this is completed, the physical layer sends the MAC layer a confirmation, PLME-SET-TRX-STATE.confirm, to indicate the result of the request.

3.2.5 Read and Write Commands

The following primitives are used to read and write PIB information:

- **PLME-GET.request:** This request is generated by the MLME and sent to the PLME to read information from the PIB.
- **PLME-GET.confirm:** The result of the PLME-GET.request containing the PIB attributes.

- **PLME-SET.request:** This is generated by the MAC layer to set attributes in the physical layer PIB.
- **PLME-SET.confirm:** This is generated by PLME to the MAC layer to notify the results of PLME-SET.request.

3.3 TRANSMITTER POWER AND RECEIVER SENSITIVITY

The transmitted power is represented in dBm (decibel with reference to 1 milliwatt [mW]). This can be calculated using the following formula, where P is the power of the transmitter in milliwatts, and X is the power in dBm:

$$X \, dBm = 10 \, log_{10} P$$

The IEEE 802.15.4 requires that output power must be at least –3dBm (0.5 mW). Reducing power transmission of the devices in a network will reduce interference with other devices.

Table 3.4 shows several values of transmission power in dBm and their corresponding power in mW.

Table 3.4   Transmission Power in dBm and Corresponding Values in Milliwatts

30 dBm	1000 mW	22 dBm	160 mW
27 dBm	500 mW	21 dBm	125 mW
26 dBm	400 mW	20 dBm	100 mW
25 dBm	316 mW	15 dBm	32 mW
24 dBm	250 mW	10 dBm	10 mW
23 dBm	200 mW	4 dBm	2.5 mW
20 dBm	100 mW	0 dBm	0.0 mW
10 dBm	10 mW	–3 dBm	0.5 mW

- **Receiver Sensitivity:** IEEE 802.14.5 defines the receiver sensitivity as the smallest receiver signal power that results in less than 1% packet error rate. The receiver sensitivity for 2.4GHz band is –85dBm and for 868/915MHz is –92 dBm.

3.3.1 Receiver Signal Strength (RSS)

According to Friis, in free-space, the power of a wireless signal at the receiver is represented by Equation 3.1.

$$P_R \, (d) = \frac{P_T \, G_T G_R \, \lambda^2}{(4\Pi)^2 d^2} \tag{3.1}$$

where the variables of the equation are defined as follows:

P_R *(d)*: Power of the signal at receiver side at distance d

   G_T: Antenna gain at transmitter side

   G_R: Antenna gain at receiver side

   λ: Wavelength of the signal where $\lambda = C / f$, C is speed of light and f frequency of the signal

   $d =$ Distance between transmitter and receiver

In Equation 3.1, the G_T, G_R, and λ are constant. Therefore, Equation 3.1 can be represented by the simpler Equation 3.2.

$$P_R (d) = \frac{KP_T}{d^2} \tag{3.2}$$

To calculate the signal power of the receiver at distance d, by experiment, the receiver power is measured at distance d_0 as a reference value and used to calculate the receiver power at distance d. The receiver power at distance d_0 is represented in Equation 3.3.

$$P_R (d0) = \frac{KP_T}{(d0)^2} \tag{3.3}$$

By using Equation 3.2 and 3.3, Equation 3.4 is produced.

$$\frac{P_R(d)}{P_R(d0)} = \frac{(d0)^2}{d^2} \tag{3.4}$$

Equation 3.4 shows that if the distance between the receiver and the transmitter is doubled, the receiver signal power is reduced by a factor of four.

3.3.2 Received Signal Strength Indication (RSSI)

In a wireless network, the RSS is converted to received signal strength indication (RSSI), represented in decibel (dBm):

$RSSI = 10\ Log\ P_R\ dBm$, where P_R is represented in milliwatts.

3.3.3 Path Loss

When a wireless signal propagates through space, it loses power. The amount of the power lost is referred to as path loss and is calculated using Equation 3.5.

$$P_L = P_T - P_R \tag{3.5}$$

Equation 3.1 can be represented by Equation 3.6 when using meters and megahertz as units of measurement:

$$P_R \ (dBm) = P_T \ (dBm) + G_T \ (dBm) + G_R \ (dBm) - 20 \ log(d) - 20 \ log(f) + 27.56 \qquad (3.6)$$

where d is in meters and f in MHz.

If the gain of the receiver and the transmitter antennas are one, the path loss can be expressed by Equation 3.7

$$_L = 20 \ Log \ (d) + 20 \ log \ (f) - 27.56 \qquad (3.7)$$

The receiver power is dependent on the frequency of the operation, distance, and environment in which the wireless device operates, such as free air, inside a building with line of sight, or inside a building without line of sight. Based on the experiment by Seidel and Rappaport, the receiver power at distance d is given by Equation 3.8.

$$P_d = P_{d0} - 10n \ log \ (f) - 10n \ log(d) + 30n - 32.44 \qquad (3.8)$$

where n is the path loss exponent, which was determined based on an experiment by Seidel and Rappaport.

Table 3.5, shows the path loss exponents for different environments.

Table 3.5   Path Loss Exponents for Different Environments

Environment	Path Loss Exponent
Free space	2
Urban area	2.7 to 3.5
Shadowed urban area	3 to 5
In-building line of sight	1.6 to 1.8
Obstructed in building	4 to 6
Obstructed in factory	2 to 3

3.3.4 Link Quality Indication (LQI)

The LQI value is used by the router to select the best route to the next hop. LQI is measured using RSSI or the signal-to-noise ratio (S/N) or the packet error rate (PER) or a combination of these methods. Each manufacturer of a transceiver uses a different method for measuring the LQI of an incoming signal. For example, Texas Instruments uses a combination of RSSI and S/N with a correlation value to indicate signal quality for the Chipcon CC240 chipset, and the ATML corporation uses PER to measure the LQI of an incoming signal. The PER is calculated by Equation 3.9:

$$PER = 1 - (1 - BER)^N \qquad (3.9)$$

where *BER* is the bit error rate.

The PER and RSSI are a function of node distance. Therefore, LQI is an estimate of link quality. LQI is represented by 8 bits, with 0 representing the lowest quality of an incoming signal and 255 representing the highest quality of an incoming signal.

The physical layer performs LQI on each packet received and passes the result to the MAC layer using the PD-Data.indication primitive. The MAC layer can then use the results for channel selection.

3.3.5 Transmission Range

The transmission range depends on the transmitter power, operating frequency, and environment in which the wireless device operates. At the receiver side, the wireless device is able to measure the received signal power and use Equation 3.5 to calculate the transmission range. Table 3.6 shows the transmission range for indoors and outdoors.

Table 3.6   Transmission Range for Indoors and Outdoors

Transmission Power	Indoor Range (Meters)	Outdoor Range (Meters)
1 mW	100	300
100 mW	300	4,000

3.4 PHYSICAL LAYER INFORMATION BASE (PIB)

The physical layer contains the Physical Layer information base (PIB). The PIB is a database that holds physical layer attributes. These attributes determine the characteristics of the physical layer for operations. Some of the physical layer attributes are constant, whereas others are variable. These attributes are represented by codes rather than words to reduce the size of the PIB.

3.4.1 PIB Constant Attributes

- **Maximum physical layer protocol data unit:** Contains 127 bytes.
- **Turnaround time for TX and RX:** The turnaround time defines the switch time between the transmitter and the receiver of the transceiver. The switching time is set to 12 symbol periods.

3.4.2 PIB Variable Attributes

The following are some of the more important PIB variable attributes:

- **Current channel:** The channel number currently used for physical layer transmission.
- **List of channels supported:** This indicates the list of channels supported by the device. The channels are represented by $B_{31} \ldots B_{27} B_{26} B_{25} \ldots B_0$, where $B_{31} \ldots B_{27}$ represents

the page number, and $B_{26} B_{25} \ldots B_0$ represents the 27 channel numbers. When B_i is logic one, the ith channel is supported by the physical layer.

- **Channel access mode:** The physical layer offers the following channel access modes:

  Mode 1: Detects the energy of the channel signal; if it is above a certain value, it indicates that the channel is busy.

  Mode 2: Used for a device that wants to access the network by sensing the channel. CCA shall report a busy medium only upon detection of a signal with the modulation and spreading characteristics of IEEE 802.15.4; signals may be above or below the energy detection threshold.

  Mode 3: Carrier sense with energy above the threshold. CCA shall report a busy medium only upon detection of a signal with the modulation and spreading characteristics of IEEE 802.15.4 and whose signal is above the energy detection threshold.

- **Current page:** Represents the current page number.

- **Symbols per byte:** Number of symbols represented in each byte.

- **Maximum and minimum duration of a frame:** Defines the maximum number of symbols and minimum number of symbols for a frame.

3.5 PHYSICAL LAYER TRANSMISSION

The physical layer uses DSSS for transmission of information. The physical layer operates in three different frequency bands: 2.4GHz, which uses O-QPSK modulation; 868MHz, which uses BPSK modulation with optional ASK and O-QPSK modulation; and 915MHz, which also uses BPSK modulation with optional ASK and O-QPSK modulation.

3.5.1 2.4GHz Transmission

A physical layer operating at 2.4GHz transmits information using O-QPSK with half sine wave pulse. In O-QPSK, each data symbol is represented by 4 bits. These 4 bits are mapped to a 32-bit chip represented by $C_0 C_1 C_2 C_3 \ldots C_{31}$. Because a symbol is represented as 4 bits, there are 16 different symbols. Each symbol is uniquely mapped to a 32-bit chip, and thus there are 16 different 32-bit chips. The symbols are transmitted at the rate of 2 million chips per second. The O-QPSK uses half sine wave for transmission of chip bits. In O-QPSK, the even chip bits modulated by a half sine wave are called in-phase (I-Phase), and odd chip bits modulated with a 90 degree shift are called Q-Phase. Assuming a chip bit $C_0 C_1 C_2 C_3 C_4 C_5 C_6 C_7$ that is equal to 11001010, Figure 3.8 shows its O-QPSK modulation.

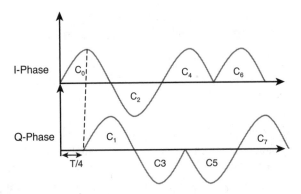

Figure 3.8 O-QPSK modulation for 11001010

The advantage of using O-QPSK is that it incurs a lower cost and results in less amplifier power consumption during transmission. Figure 3.9 shows the 2.4GHz modulation process where the 4-bit data symbol is mapped to the 32-bit chip, and the chip is modulated through O-QPSK modulation.

Figure 3.9 2.4GHz modulation process

3.5.2 868/915MHz BPSK

The 868 and 915MHz bands use DSSS, where each bit is mapped to a 15 chip bits. Binary phase-shift keying is then applied to each chip bit for transmission. Table 3.7 shows the chip bits for the 868/915MHz bands.

Table 3.7 Chip Bits for DSSS

Bit	Chip Value
0	111010110001000
1	00010100110111

BPSK uses two phases to represent logic 0 and 1. The two phases are separated by 180 degrees, as shown in Figure 3.10.

Figure 3.11 shows the modulation process. Each bit of the PPDU is encoded using a differential encoder; each encoded bit is then converted to its 15-bit chip bit. Lastly, each chip bit is modulated using BPSK.

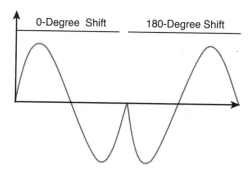

Figure 3.10   Binary-shift keying

Figure 3.11   868/915MHz modulation process

3.5.3 Differential Encoding

The purpose of differential encoding is to protect the bit stream from polarity reversal. Figure 3.12 shows BPSK. In BPSK, the receiver cannot detect the phase of the incoming signal, whether it is shifted 180 degrees or not. In Figure 3.12, the receiver reads the first two cycles as 00 rather than 11. Then, at the point of phase reversal, the receiver reads the last 2 bits as 11 rather than 00. That is, the receiver reads 1100 rather than 0011; this is called phase ambiguity. The phase ambiguity can be corrected by using deferential encoding.

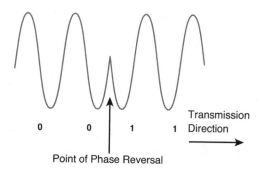

Figure 3.12   Binary phase-shift keying

Figure 3.13 shows a differential encoder at the transmitter side. On the transmitter's side, the bit stream is supplied to the differential encoder and, on the receiver's side, the encoded bit stream passes through the differential decoder.

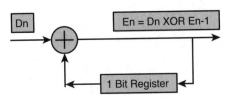

Figure 3.13   Differential encoder

The En-1 can be chosen to be either 1 or 0 (initial value of shift register). In Figure 3.14, the En-1 is set to 0, and the input to the encoder is set to 110011, which results in an output of 010001. This is produced by performing an XOR operation on the rightmost bit of the input and the En-1 bit. This XOR operation is repeated using the next bit in the input and the result of the previous XOR operation until the input is exhausted.

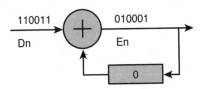

Figure 3.14   Differential encoder process

The result of the encoding, 010001, is transmitted to the receiver, where a differential decoder is used to recover the original bit stream. Figure 3.15 shows the differential decoder and the differential decoder process for the input 010001

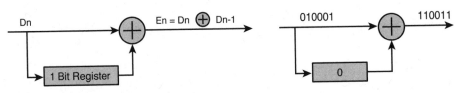

Figure 3.15   Differential decoder and decoder process for input 010001

If the polarity of the bit stream reverses and 101110 is received instead of 010001, the differential decoder can recover the original bit stream. Figure 3.16 shows this process.

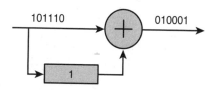

Figure 3.16   Differential decoder process for input 101110

3.5.4 868/915MHz Optional Modulation

The 868/915MHz bands optionally use parallel sequence-spread spectrum (PSSS) with ASK and O-QPSK modulation. In ASK, the amplitude of the signal changes. This is also referred to as amplitude modulation (AM). The receiver recognizes these modulation changes as voltage changes, shown in Figure 3.17. The smaller amplitude is represented by *0* and the larger amplitude is represented by *1*. It can also represent the original signal with more than two amplitudes. Each cycle is represented by 1 or more bits.

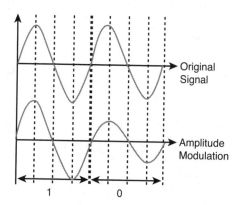

Figure 3.17   ASK

Parallel Sequence-Spread Spectrum

PSSS uses a 32-bit chip sequence: 20-chip sequences for the 868MHz band, and 5-chip sequences for the 915MHz band. Table 3.8 shows the chip bits for PSSS using the 5-chip sequences in their bipolar level form.

Table 3.8 Chip Bits for Parallel Sequence-Spread Spectrum

Bit Number	Chip Sequence 0	Chip Sequence 1	Chip Sequence 2	Chip Sequence 3	Chip Sequence 4
0	−1	1	−1	1	1
1	−1	1	−1	1	−1
2	−1	−1	1	1	1
3	−1	1	1	1	1
4	1	−1	−1	1	−1
5	−1	1	1	−1	−1
6	−1	−1	1	−1	1
7	1	−1	1	−1	1
8	1	1	−1	−1	−1
9	1	−1	−1	−1	−1
10	1	1	−1	−1	1
11	−1	−1	1	1	−1
12	−1	−1	−1	1	−1
13	1	1	1	−1	1
14	1	−1	−1	1	1
15	1	1	−1	1	1
16	1	1	1	−1	−1
17	1	−1	−1	1	1
18	−1	−1	−1	−1	1
19	−1	1	−1	−1	−1
20	−1	1	−1	−1	−1
21	1	1	1	−1	−1
22	1	1	1	1	−1
23	−1	1	1	−1	1
24	1	−1	−1	−1	−1
25	1	−1	1	1	−1
26	1	−1	1	−1	−1
27	−1	1	1	1	−1
28	1	1	1	1	1
29	−1	−1	1	−1	−1
30	1	1	−1	−1	−1
31	−1	1	−1	1	1

Because of the similarities between the 868 and 915MHz bands' optional coding, only the 915MHz band's optional coding is examined in detail.

915MHz Band Optional Coding

The following steps detail the PSSS operation for the 915MHz band:

1. The PPDU bits are converted to bipolar level, meaning 0 is represented by −1 and 1 is represented by +1.

2. The PPDU is broken into segments, and each segment is represented by 5 bits: $B_4 B_3 B_2 B_1 B_0$ and each bit converted to bipolar representation (0 to −1 and 1 to +1). Each bipolar bit of the segment is multiplied by the corresponding chip sequence listed in Table 3.7.

B_0 * 32-bit Chip sequence0

B_1 * 32-bit Chip sequence1

B_2 * 32-bit Chip sequence2

B_3 * 32-bit Chip sequence3

B_4 * 32-bit Chip sequence4

3. The results of the multiplications are added together.

4. The preamble portion of the PPDU is transmitted using BPSK, and the remaining PPDU (SFD, frame length, and PSDU) are transmitted using ASK.

Figure 3.18 shows the entire PSSS process.

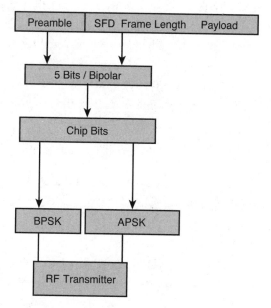

Figure 3.18   PSSS process for 915MHz

Any type of modulation can be applied to PSSS (for example, QAM or OFDM). The key characteristics of PSSS are

- Better response on multipath fading
- Incurs less bit error
- Supports high data rate
- Low energy consumption
- Longer range

SUMMARY

This chapter presented the IEEE 802.15.4 physical layer. The following are the main points covered in the chapter.

- The IEEE 802.15.4 physical layer operates in the 868/915MHz and the 2.4GHz ranges of the ISM band and, in the European countries, operates in the 868MHz band.
- The 868/915MHz bands use BPSK modulation and optionally support O-QPSK and ASK modulations.
- The physical layer offers three pages for channel selection: page 0, which contains 27 channels; and pages 1 and 2, which contain 11 channels each.
- IEEE 802.14.5 defines the receiver sensitivity as the smallest receiver signal power that results in less than 1% packet error rate.
- The receiver sensitivity for 2.4MHz band is –84 dBm and for 868/915Mhz is –92 dBm.
- The transmission range depends on the transmitter power, operating frequency, and the environment in which the wireless device operates.
- The link quality indication (LQI) value is used by the router to select the best route to the next hop.
- LQI is measured using RSSI or signal-to-noise ratio (S/N), packet error rate (PER), or a combination of these methods.
- Differential encoding is used to overcome the ambiguity of a binary phase shift.
- The physical layer provides services to the MAC layer and transceiver.
- The physical layer contains a database known as the PIB, which holds physical layer constant and variable attributes.
- The function of the physical layer is to provide data transfer and management services. The management services are accomplished by the physical layer management entity (PLME).
- The physical layer uses the PD-Data.request, PD-Data.confirm, and PD-Data.indication primitives for data transfer.
- The physical layer can perform three types of channel scanning: active, passive, and orphan scans.
- The maximum payload of the physical layer protocol data unit (PPDU) is 127 bytes.
- The PLME performs energy detection (ED), link quality indication (LQI), channel frequency selection, and clear channel assessment (CCA).
- The 2.4GHz band uses O-QPSK modulation.
- 868/915MHz bands use differential encoding to correct bit stream errors due to polarity reversal.

REFERENCES

1. IEEE 802.15.4 Specification 2003

2. www.cirronet.com/pdf/wp_ZigBeePowerOptions.pdf

3. Shreharsha Rao, Estimating the ZigBee Transmission Range ISM Band, EDN, May 2007

4. The Institute of Electrical and Electronics Engineers Inc. Press, *Wireless Communications Principles and Practice,* at 104 (1996)

5. Pahllavan, K. and Krishnamurthy, P. *Principles of Wireless Networks,* Prentice-Hall, 2002

6. Elahi and Elahi, *Data, Network &Communications Technology,* Thomson 2006

7. Farahani, S. ZigBee Wireless Network and Transceivers, Newnes, 2008

8. Gutierrez, J., Gallaway, E., and Barrett, R. *Low-Rate Wireless Personnel Area Networks,* IEEE Press Publication, 2007

9. Alberola, R. and Pesch, D. Extending Avorora with an IEEE 802.15.4 Compliant Radio Chip model, Cork Institute of Technology, Ireland10. Seidel and Rappaport, "914 MHz Path-Loss Prediction Models for Indoor Wireless Communications in Multi-Floored Building," IEEE Transaction on Antenna and Propagation, Feb. 1992

CHAPTER 4

IEEE 802.15.4 MEDIA ACCESS CONTROL (MAC) LAYER

INTRODUCTION

The Media Access Control (MAC) layer, specified by IEEE 802.15.4, is located between the physical and network layers. The MAC layer performs the following functions:

- Transfers data to the network layer and vice versa; transfers data to the physical layer and vice versa
- End device association and disassociation
- In the coordinator, offers optional guaranteed time slot (GTS) for each device accessing the network
- Generates the beacon frame in a coordinator
- Supports device security
- Provides carrier-sense multiple access with collision avoidance (CSMA/CA) as the access method for the network
- Provides a reliable connection between two MAC layers by using an acknowledgment frame

4.1 MAC LAYER SERVICES

Figure 4.1 shows the MAC layer architecture and its interface to the network and physical layers. The MAC layer's functionality is divided into two parts: data transfer, which is performed by the MAC common part sublayer

(MCPS); and management functions, which are performed by the MAC layer management entity (MLME). The function of the MCPS is to transfer data from the network layer to the MAC layer and vice versa. It also transfers data packets from the MAC layer to the physical layer and back. The function of the MLME is to manage the MAC layer. The MAC layer also contains a MAC information base (MIB) (database) that holds attributes relevant to the MAC layer.

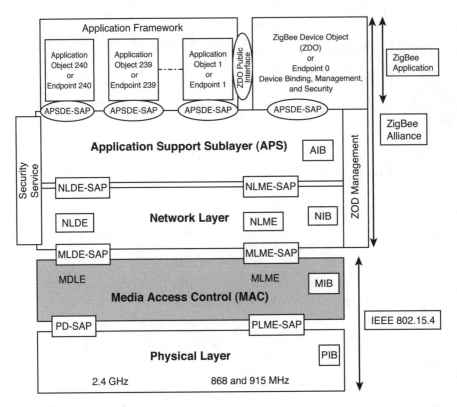

Figure 4.1 MAC layer architecture

4.1.1 MAC Data Services

The MCPS provides data services by using the following primitives:

- MCPS-DATA.Request
- MCPS-DATA.Confirm
- MCPS-DATA.Indication
- MCPS-PURGE.Request
- MCPS-PURGE.Confirm

MCPS-DATA.Request: This primitive is generated by the network layer and issued to the MAC layer to indicate the transfer of the network protocol data unit (NPDU) to a destination MAC layer. The MAC layer adds its MAC header information to the NPDU, and the resulting packet is called the MAC protocol data unit (MPDU). The MAC layer then sends the MPDU to the physical layer, where the physical layer adds its header information, forming a packet called the physical layer protocol data unit (PPDU). And then, the physical layer transmits the PPDU to the destination. Figure 4.2 shows MCPS-DATA.Request format.

Source Address Mode	Source PANID	Source Address	Dest Address Mode	Dest PANID	Dest Address	msdu Length	msdu	msdu Handle	TX Option	Security Level	Key ID Mode	Key Source	Key Index

Figure 4.2   MCPS-DATA.Request format

Each node is assigned a unique address which can either be a 64-bit address assigned by IEEE or a 16-bit short address. Additionally, each network has its own address known as the Personal Wireless Network ID (PAN ID) which is represented by 16 bits.

- **Source address mode:**
  - 00 Source does not have an address.
  - 01 Reserved.
  - 02 Source address is 16 bits.
  - 03 Source address is 64 bits.
- **Source PAN ID:** 16-bit address used for the PAN ID.
- **Source address:** Represents the address of the source node.
- **Destination address mode:**
  - 00 Destination does not have an address.
  - 01 Reserved.
  - 02 Destination address is 16 bits.
  - 03 Destination address is 64 bits.
- **Destination PAN ID:** 16-bit address used for the PAN ID.
- **Destination address:** Destination node address.
- **MSDU length:** Defines the number of bytes in the MSDU.
- **MSDU:** The MAC layer service data unit or MAC payload.
- **MSDU handle:** Identification for the MSDU.
- **Tx option:** The tx option is represented by 3 bits, $b_2 b_1 b_0$, where
  if $b_0 = 1$, source requests acknowledgment from destination.
  if $b_1 = 1$, $b_1 = 0$, the MAC layer uses Contention Access Period.

If $b_2 = 1$, indirect addressing for transmission is to be used; that is, there will be no source address in the frame.

- **Security level:** Contains security levels ranging from 00 to 07. When set to 0, the MAC layer will not apply security to the MAC payload; otherwise, security is applied to the payload. For greater detail regarding the various security levels, refer to Table 8.1 in Chapter 8, "ZigBee Security."

- **Key ID mode:** Indicates the mode under which the key is generated. For example, it can indicate whether the key is generated by the source or the destination.

- **Key source:** Defines the source for which the key will be used.

- **Key index:** Defines the index of the key.

- **MCPS-DATA.Confirm:** This is generated by the MAC layer and sent to the network layer in the response to a MCPS-DATA.Request. Figure 4.3 shows the MCPS-DATA.Confirm primitive.

Figure 4.3   MCPS-DATA.Confirm

- **MSDU handle:** Identification for the MSDU. Matches the MSDU handle, which was sent in the request.

- **Status:** Store the status of the request: success, invalid address, or frame too long.

- **Time stamp:** The time that the primitive was sent. (Optional)

- **MCPS-DATA.Indication:** This is generated by the MAC layer to the network layer for the purposes of indicating the transfer of the MSDU (MAC service data unit). Figure 4.4 shows the MCPS-DATA.Indication primitive.

Source Address Mode	Source PANID	Source Address	Dest Address Mode	Destination PANID	Destination Address	msdu Length	msdu	Link Quality	DSN	Time Stamp	Security Level	Key ID Mode	Key Source	Key Index

Figure 4.4   MCPS-DATA.Indication

The fields contained in the MCPS-DATA.Indication primitive are similar in format and function to those depicted in Figure 4.2, with the exception of the following:

- **LQI (link quality indication):** Indicates the quality of the channel during reception of a frame; the LQI is between 00h to FFh, where lower numbers represent lower link quality.

- **DSN (data sequence number):** Sequence number of frame.
- **Time Stamp:** The time that the frame was received. (Optional)

Figure 4.5 shows the MAC layer data transfer process. The process begins with the transmission of the MCPS-DATA.Request from the network layer to the MAC layer. The MAC layer then transmits the PD-DATA.Request containing the MPDU to the physical layer. When the physical layer transmission is completed, it issues a PD-DATA.Confirm to the MAC layer. If the source MAC layer had set b_0 of the TX option field to 1, the MAC layer expects to receive an ACK packet from the destination. It will wait for this ACK packet for a time no greater than the ACK wait duration specified in the MIB. If the source does not receive an ACK frame, it will attempt to retransmit the frame; the number of additional attempts is specified in the MIB as the maximum frame retries value. However, if the ACK frame is not received after the maximum number of attempts has been reached, the upper layer is informed that the destination is unreachable.

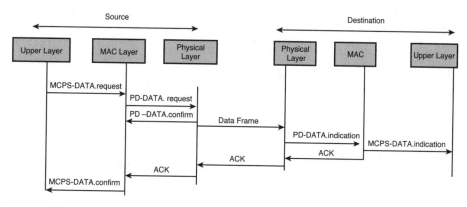

Figure 4.5   MCPS-DATA.Request operation

- **MCPS-PURGE.Request:** This is generated by a higher layer and is made to the MAC layer to purge a MAC service data unit from the MAC queue.
- **MCPS-PURGE.Response:** This is generated by the MAC layer in response to a MCPS-PURGE.Request to indicate the result.

4.2 MAC LAYER INFORMATION BASE (MIB)

The MAC Layer information base (MIB) contains the attributes necessary for managing a device's MAC sublayer. Like the Physical Layer information base (PIB), these attributes are represented by codes rather than by actual words and can be either constant or variable.

4.2.1 MIB Constant Attributes (Read–Only)

- **Extended Address:** 64 bit IEEE address.

- **MAC minimum short interframe spacing:** Defines the space between two consecutive short frames as represented by a number of symbols. It allows time for the receiver to process a frame before the next frame is received.

- **MAC minimum long interframe spacing:** Similar to the short interframe spacing, it defines the space between two consecutive long frames.

- **Base slot duration:** The number of symbols represented in one slot of the superframe. It is set to 60 symbols.

- **aMaxBE:** The maximum value of the backoff algorithm used in carrier-sense multiple access with collision avoidance (CSMA/CA). Set to 5.

- **aMax Frame retries:** The number of times a frame can be retransmitted due to a transmission failure. It is set to 3.

- **aMax MAC frame size:** The maximum number of bytes in a MAC frame that can be followed by a short interframe space (SIFS). It is set 18.

- **Number of slots in superframe:** 16 slots.

- **Unit backoff period:** 60 symbols.

4.2.2 MIB Variable Attributes

The MIB variable attributes (or MAC PIB) are used in the management of the MAC sublayer. The MIB contains 32 attributes, each of which is represented by a code. Table 4.1 lists some of the MIB attributes and their codes and descriptions.

Table 4.1 MIB Attributes

Attribute Name	Code (Hex)	Range	Description
MAC ACK Wait Duration	40	54–120 default 54	Maximum number of symbols for which the source will wait for the ACK frame.
MAC Association PAN Coordinator	56	True/false	True indicates the device is associated with the PAN through the coordinator.
MAC Association Permit	41	True/false	True: The coordinator accepts new-association. False: The coordinator does not accept association.
MAC Auto Request	42	True/false	True: The device automatically sends data requests to the coordinator if the device address is in the beacon frame.
MAC Beacon Order (BO)	47	0–15	Defines the beacon interval; when BO=15, the coordinator will not transmit the beacon.

Table 4.1 Continued

Attribute Name	Code (Hex)	Range	Description
MAC Coordinator Extended Address	4a	64 bits	64-bit IEEE address.
MAC Coordinator Short Address	4b	16 bits	16-bit short address.
MAC Max CSMA/CA Backoffs	4e	0–5, default 4	Number of times the CSMA/CA attempts to access the network.
MAC Min BE	4f	0–3, default 3	The minimum backoff value for the CSMA/CA exponent.
MAC PANID	50	16 bits	PAN address.
MAC frame Retries	59	0-7	Number of times a frame can be retransmitted because of a failure.
MAC Short Address	53	0000–FFFF	Short address of device. FFFE means the device is associated but has not been assigned an address. FFFF means the device does not have a short address.
MAC Security Enable	5D	True/false	True means MAC security is enabled, and false means the MAC security is disabled.
MAC Time Stamp	5C	True/false	True means the MAC layer supports a time stamp; false means the MAC layer does not support a time stamp.

4.3 MAC MANAGEMENT SERVICES

The MAC layer management entity (MLME) is responsible for MAC management; that is, it manages the operation of the MAC layer through a collection of primitives. The MLME request primitives are generated by the network layer and transmitted to the MAC layer of the device. Each primitive, listed here, has a corresponding confirm primitive.

- **MLME-BEACON-NOTIFY.Indication:** When the MAC layer of a device receives a beacon frame, it sends the beacon parameters to the higher layer.
- **MLME-ASSOCIATE.Request:** It is generated by the network layer and sent to the MAC layer. It is used by an unassociated device that is requesting association with the router or coordinator.
- **MLME-ASSOCIATE.Indication:** This is generated by the MAC layer and sent to the network layer of the coordinator to indicate the reception of MLME-ASSOCIATE.Request.

- **MLME_ASSOCIATE.Response:** This is generated by the network layer and sent to the MAC layer of the coordinator in the response to an MLME-ASSOCIATE.Indication. The MLME of the coordinator then generates an associate response command for the device requesting association.

- **MLME-DISASSOCIATE.Request:** Used by an associated device to request disassociation from the network. It can also be used by the coordinator to request that an associated device leave the network.

- **MLME-GET.Request:** Used by a device to read the MIB information.

- **MLME-GTS.Request:** A device request for the coordinator to allocate a new guaranteed time slot (GTS) or deallocate the current GTS

- **MLME-ORPHAN.Indication:** When the MAC layer of the coordinator receives an orphan notification command, the MAC layer informs the network layer by sending the MLME-ORPHAN.Indication.

- **MLME-RESET.Request:** Used to set the MAC PIB attributes to their default values.

- **MLME-SCAN.Request:** Used to scan a given list of channels.

- **MLME-SET.Request:** Used to write a value in the MIB.

- **MLME-SYNC.Request:** Indicates a device requesting a beacon from the coordinator for synchronization.

- **MLME-SYNC-LOST.Request:** Used by a device to synchronize with the coordinator.

- **MLME-POLL.Request:** A device sends this request to the coordinator to retrieve any waiting data.

- **MLME-RX-ENABLE:** Enables the receiver of the transceiver.

- **MLME-START:** Starts a network.

4.3.1 MLME–ASSOCIATE.Request

This primitive is generated by a higher layer (network layer) and sent to the MAC layer to request association with a coordinator or full-feature device (FFD). The MAC layer of the device transmits an MLME-ASSOCIATE.Request to the MAC layer of the coordinator. Then the MAC layer of the coordinator sends an MLME-ASSOCIATE.Indication to the network layer. Having received the association indication, the coordinator's network layer makes the decision to either accept this new device or reject it. This decision is communicated to the device MAC layer with an MLME-ASSOCIATE.Response. The device MAC layer sends an MLME-ASSOCIATE.Confirm to its network layer using the information contained within the coordinator association response to indicate the association status (for example, success or failure). Figure 4.6 shows the association process, and Figure 4.7 shows the MLME-ASSOCIATE.Request primitive.

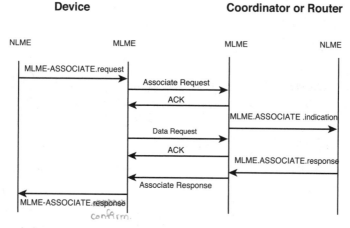

Figure 4.6   Association request process

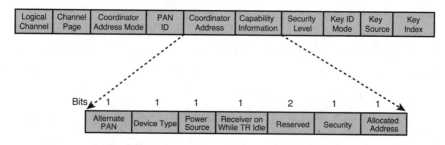

Figure 4.7   MLME-ASSOCIATE.Request primitive

- **Logical channel:** The channel number that a device is using for association.
- **Channel page:** The page number on which the channel is located.
- **Coordinator address mode:** Uses a 16-bit short address or a 64-bit IEEE address.
- **Coordinator PAN ID:** The ID of the network in which a coordinator is located. It is represented as 16 bits.
- **Coordinator address:** The address of the coordinator.
- **Capability information fields:** Contains information about the end device requesting association with the coordinator or router. This information includes:

  Alternate PAN: If this bit is set to 1, the device is capable of becoming the coordinator.

  Device type: This bit, when set to 1, indicates that the device is an FFD; otherwise it is considered a reduced-function device (RFD).

  Power source: When set to 1, the device is using main power; otherwise it is using a rechargeable battery.

Receiver on while transceiver is idle: If this bit is set to 1, the receiver of the end device is on while the transceiver is idle; otherwise the receiver is off.

Security capability: This bit, when set to 1, indicates that the end device is able to send and receive secured MAC frames.

Allocation address: When this bit is set to 1, the coordinator is requested to allocate an address to the end device.

4.3.2 MLME-ASSOCIATE.Confirm

This is generated by the coordinator to inform a device of its association status. The primitive is shown in Figure 4.8.

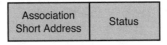

Figure 4.8 MLME-ASSOCIATE.Confirm

- **Association short address:** The coordinator assigned a 16-bit address to a device.
- **Status:** Indicates the status of an association attempt. It contains success, channel access failure, no acknowledgment (ACK), or failed security check. A failed security check indicates the device is not on the access control list (ACL).

4.3.3 MLME-ASSOCIATE.Indication

When a coordinator's MAC layer receives an association request from a device, the MAC layer sends an MLME-ASSOCIATE.Indication to its network layer. Figure 4.9 shows this MLME-ASSOCIATE.Indication primitive. After the coordinator has decided to accept or reject the device, it responds with an MLME-ASSOCIATE.Response, as shown in Figure 4.10.

Device Address	Associate Short Address	Status	Security Level	Key ID Mode	Key Source	Key Index

Figure 4.9 MLME-ASSOCIATE.Indication

Device Address	Associate Short Address	Status	Security Level	Key ID Mode	Key Source	Key Index

Figure 4.10 MLME-ASSOCIATE.Response

- **Device address:** 64-bit IEEE address.
- **Associate short address:** 16-bit device address.
- **Status:** Indicates the association status: success, access denied, and PAN does not have capacity.

4.3.4 MLME-BEACON-NOTIFY.Indication

When the MAC layer receives a beacon frame, it uses MLME-BEACON-NOTIFY.Indication to send the beacon's parameters to the network layer; Figure 4.11 shows the MLME-BEACON-NOTIFICATION.Indication primitive.

BSN	PAN Descriptor	PAN Address	Address List	SDU Length	SDU

Figure 4.11   MLME-BEACON-NOTIFY.Indication

- **BSN (beacon sequence number):** Any number from 0x00 to 0xFF.
- **PAN Descriptor:** This field contains the following information about the network:

  Coordinator address mode (indicates whether the coordinator uses 16-bit or 64-bit addresses)

  Coordinator PAN ID, represented as 16 bits

  Coordinator address

  Current channel

  Superframe specification, if any

  Link quality of beacon frame received

  Security indication, if security is applied to the beacon frame
- **PAN address spaces:** Defines the number of addresses in the address list.
- **Address list:** List of end device addresses for which the coordinator currently holds data.
- **SDU length:** Number of bytes in the SDU.
- **SDU:** Payload of the beacon frame.

4.3.5 MLME-SCAN.Request

This is generated by the network layer to request that the MAC layer scan a list of channels. This primitive, represented in Figure 4.12, is used for active scan, orphan scan, passive scan, and energy detection. When a scan is completed, the MAC layer sends an MLME-SCAN.Confirm to the network layer.

Scan Type	Scan Channels	Scan Duration	Channel Page	Key ID Mode	Key Source	Key Index

Figure 4.12   MLME-SCAN.Request

- **Scan type:**

  00   Energy detection used only by FFD

  01   Active scan used by FFD

  02   Passive scan

  03   Orphan scan
- **Scan channels:** A list of channels to be scanned.
- **Scan duration:** Defines the duration of the scan for each channel.
- **Energy detection:** Energy detection is used by a device to measure the energy level of the selected channel. This can be used by a coordinator to select the best channel. The network layer sends an MLME-SCAN.Request to the MAC layer by setting the scan type to 00.

4.3.6. MLME–START.Request

This is generated by the network layer and issued to the MAC layer for the coordinator to generate a new superframe. This is called after the network layer has completed an NLME-NETWORK-FORMATION.Request. The NLME of the network layer then sends an MLME-START.Request to start the network. Figure 4.13 shows this MLME-START. Request primitive.

Coordinator PANID	Logical Channel	Beacon Order	Super-Frame Order	PAN Coordinator	Battery Life Extension	Coordinator Realignment	Security Enable

Figure 4.13   MLME-START.Request

- **Coordinator PAN ID:** This is the network ID (16 bits).
- **Logical channel:** The channel number at which the coordinator starts transmitting.
- **Beacon order (BO):** A value usually between 0 and 14. If the value is set to 15, the coordinator does not transmit a beacon.
- **Superframe order (SO):** This value is between 0 and 14 and is used to define the superframe active portion.
- **PAN coordinator (true/false):** A value of true indicates the device will become a coordinator; false means it will not be a coordinator and may be a router.

- **Battery life extension (true/false):** A value of true indicates that the receiver of the coordinator is disabled for the duration of full backoff periods after the interframe spacing (IFS); false means that the receiver of the coordinator transmitting the beacon frame is enabled during the contention access period (CAP).

- **Coordinator realignment (true/false):** A value of true indicates that the coordinator will send a realignment command to its devices before changing its superframe.

- **Security enable (true/false):** A value of true enables security on the beacon; a false indicates security is disabled on the beacon frame.

4.3.7 MLME-START.Confirm

This primitive is generated by the MAC layer to indicate to the network layer the status of the MLME-START.Request. Status values include success, frame too long, failed security check, and invalid parameters.

4.3.8 MLME–POLL.Request

This is generated by the network layer and sent to the MAC layer of the end device requesting data from the coordinator. Figure 4.14 show the MLME-POLL.Request primitive.

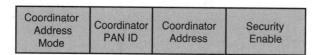

| Coordinator Address Mode | Coordinator PAN ID | Coordinator Address | Security Enable |

Figure 4.14   MLME-POLL.Request primitive

- **Coordinator address mode:** Indicates whether the coordinator uses 16-bit or 64-bit addresses

- **Coordinator PAN ID:** The 16-bit network ID

- **Coordinator Address:** Either 16 or 64 bits, depending on the value of the coordinator address mode

4.4 SCANNING CHANNELS

One function of the MAC layer is to scan channels to determine channel energy level. There are three types of scans whose availability is contingent on the type of device performing the scan. A reduced-function device (RFD) is only able to perform orphan or passive scans, whereas a full-function device (FFD) (coordinator or router) can perform active scans and energy detection on selected channels:

- **Active scan:** The active scan is used by an FFD to determine whether any network is located in its vicinity. This is done by sending a beacon request command. The network layer sends an MLME-SCAN.Request to the MAC layer by setting the scan type to 01 (active scan).

  The FFD (not just the coordinator) uses active scanning to determine if there is a device that is configured to transmit beacon frames within its vicinity. The device transmits a beacon request command, to which any such device within range will respond. The MAC layer discards those frames it receives during the scan operation that are not beacon frames. Before performing an active scan, the MAC layer stores the current PAN ID and then sets the PAN ID to 0xffff in order to accept all beacon frames. It will then restore the PAN ID.

- **Passive scan:** The network layer sends an MLME-SCAN.Request to the MAC layer by setting scan type to 02 (passive scan). The FFD or RFD listens for any beacon transmitted by any coordinator. During the scan (listening), the MAC layer will only accept beacon frames and discard all other frames. Like active scan, before performing the scan, the MAC layer stores the current PAN ID and sets the PAN ID to 0xffff to accept all beacon frames. It will then restore the PAN ID.

- **Orphan scan:** Orphan scan is used by a device that has lost its parent and is trying to reassociate itself with the parent device. During the scan process, the MAC layer discards all frames except for the realignment MAC command.

4.5 ACCESS METHOD

The MAC layer defines two methods for a device to access the network: the beacon-enabled access method, also called slotted CSMA/CA; and the non-beacon-enabled access method, also called unslotted CSMA/CA.

4.5.1 Superframe

The beacon-enabled access method uses the superframe, which is broadcasted by the coordinator to the devices. As shown in Figure 4.15, the superframe consists of 16 equal time slots. The first and last time slots contain the beacon frame. The active portion of the superframe is called the contention access period (CAP). In slotted CSMA/CA, the device should access the network during the CAP and complete it transmission within the CAP window. The superframe can have an active and an inactive period, as shown in Figure 4.16.

Some devices require a specific bandwidth for transferring information. To accommodate this, the superframe can allocate a GTS to these devices during the contention-free period (CFP), as shown in Figure 4.17. The CFP is used by devices that require low latency for transmission. However, ZigBee did not adapt GTS in its stack.

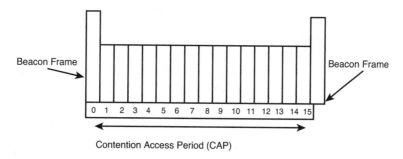

Figure 4.15 Superframe structure with CAP

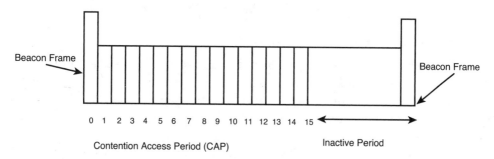

Figure 4.16 Superframe structure with inactive period

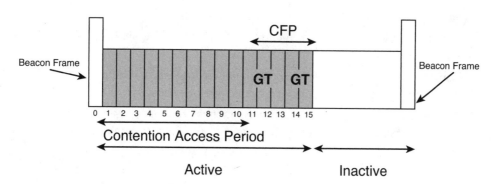

Figure 4.17 Superframe architecture

Size of superframe: The superframe size, represented by a number of symbols, defines the duration of the superframe. The superframe duration (SD) is derived using Equation 4.1, where the number of slots in a superframe can be at most 16, the base slot duration is 60 symbols, and the superframe order (SO) determines the length of the active portion of the superframe. SO values range from 0 to 15; however a value of 15 indicates that the superframe should be ignored.

$$SD = number\ of\ slots\ *\ base\ slot\ duration\ *\ 2^{SO}\ symbols \qquad (4.1)$$

Therefore, if it is assumed that the base slot duration is 60 and the number of slots is 16, $SD = 16 * 60 * 2^{SO}$ symbols.

Beacon interval (BI): Defines the interval at which the beacons are transmitted by a coordinator. It is calculated using the Equation 4.2, where the beacon order (BO) ranges from 0 to 15. Again a value of 15 indicates that the coordinator should not transmit a beacon.

$$Beacon\ Interval\ (BI) = superframe\ duration * 2^{BO} \tag{4.2}$$

4.5.2 Non-Beacon-Enabled Access Method

In this method, the coordinator does not transmit a beacon until it receives a beacon request from a device. Devices use unslotted CSMA/CA to access the network.

4.5.3 Beacon-Enabled Access Method

In a beacon-enabled network, the coordinator broadcasts the superframe periodically. Any device wanting to access the network must access the network during the superframe boundaries by using CSMA/CA. Before we look at the operations of CSMA/CA, it is necessary to define the following terms:

- **Unit of backoff algorithm:** Defined by 60 symbols, it is the time that a device must wait before accessing the network again.
- **macMaxBE:** Maximum backoff exponent, always 5.
- **macMinBE:** Minimum backoff exponent, always 3.
- **Number of backoffs (NB):** The number of times a device via CSMA/CA has tried to access the network.
- **Contention window (CW):** The contention window length is defined as the number of backoff periods that a channel must be clear before transmission can proceed.
- **Backoff exponent (BE):** When a device attempts to transmit information over a busy channel, it will wait (back off) so that it may attempt to retransmit over a clear channel.
- **MacBattery life extension (BLE) (true/false):** When this value is set to true, a device using CSMA/CA during the CAP will use the MAC battery life extension period (MAC IB attributes) to calculate the number of backoff periods.
- **MacMaxCSMABackoffs:** The maximum number of times an end device can perform CSMA/CA to access a network; the default value is 4.
- **Backoff period boundary:** The start of the beacon in the superframe.

4.5.4 CSMA/CA Flowchart

The following steps describe how a device uses CSMA/CA for accessing the network. Figure 4.18 shows a visual representation of these steps:

1. Check whether the PAN uses slotted CSMA/CA (superframe) or unslotted (no superframe). If the PAN uses slotted CSMA/CA, start from Step 2; otherwise, start from unslotted CSMA/CA.

2. Initialize NB = 0 and CW = 2.

3. Check the coordinator's battery life extension.
   If false, BE = MacMinBE = 3.

   If true, BE = Min(2, MacMinBE) = 2. This is used to ensure that the device has enough battery life to complete the CSMA/CA.

4. Find the start of the beacon frame.

5. Wait for $(2^{BE} - 1)$ * unit backoff period or $(2^{BE} - 1)$ * 60 periods.

6. Perform clear channel assessment (CCA).

7. If the channel is clear, go to Step 11; otherwise, go to Step 8.

8. Initialize CW = 2, NB = NB + 1 and BE = Min(BE + 1, MacMaxBE).

9. If NB > MacMaxCSMABackoffs, go to Step 10; otherwise, repeat from step 5.

10. Channel access failed.

11. CW = CW − 1.

12. If CW = 0, go to 13; otherwise, go to 6.

13. Success; device is ready to transmit frame.

Unslotted CSMA/CA: There is no superframe used in the unslotted CSMA/CA. Accessing the network via unslotted CSMA/CA follows the below process and can be seen visually in figure 4.16.

1. Set NB = 0, and BE = MacMinBE.

2. Wait for $(2^{BE} - 1)$ * unit backoff period or $(2^{BE} - 1)$ * 60 periods.

3. Perform CCA.

4. If channel is clear, go to 8; otherwise, go to 5.

5. Initialize NB = NB + 1, BE = Min(BE + 1, MacMaxBE).

6. If NB > MacMaxCSMABackoffs, go to 7; otherwise, go to 2.

7. Channel access failed.

8. Success.

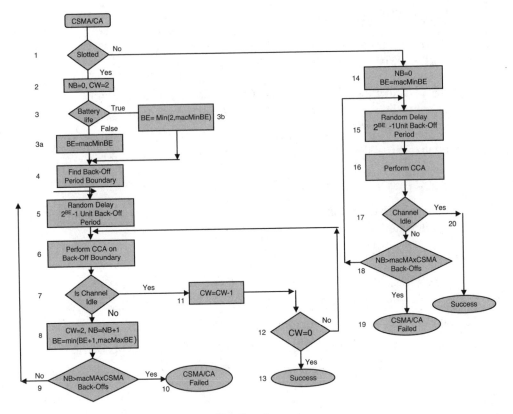

Figure 4.18 Slotted and unslotted CSMA/CA flowchart

4.6 DATA TRANSFER MODEL

ZigBee applications support multiple types of data traffic. The types of traffic it supports are as follows:

- **Periodic data traffic:** This type of data traffic is generally exhibited in electrical, water, and gas meters that need to transmit their information at specific times. A Zig-Bee node can wake up at a specific time and transmit its information and return to sleep mode.

- **Intermittent data traffic:** This type of data traffic is generated by wireless switches that are toggled on or off (for example, to control a light). When a switch position changes, the wireless device is associated with a network and transmits the switch position to the network.

- **Repetitive low-latency data traffic:** This is generated by applications that require quality of service (QoS). This is necessary in wireless devices that monitor patient conditions or security systems.

4.6.1 Data Transfer from Coordinator to Device

The data transfer command is sent by a device to the coordinator in the following cases:

- In a beacon-enabled network, setting the MAC auto request attribute in the MIB to true will cause the device to send data transfer requests automatically to the coordinator if its address is listed in the beacon frame.
- The data request command is issued when the MAC layer receives an MLME-POLL.Request from an upper layer.
- During the association process, the data request command is sent from the device to the coordinator, as shown in Figure 4.19.
- Data transfer from the coordinator to a device in a beacon-enabled network:

  The coordinator indicates in its beacon that data is pending for the device.

  The device listens to the beacon, checking whether data is pending for that device.

  The device uses slotted CSMA/CA to send a data request to the coordinator.

  The coordinator responds to the data request with an ACK frame.

  The coordinator transfers data to the device, as shown in Figure 4.19.

  The device sends an ACK frame to the coordinator when the transfer has been completed.

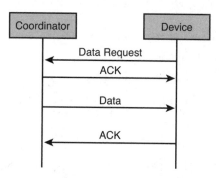

Figure 4.19   Data transfer from a coordinator to a device in a beacon enabled network

- Data transfer from the coordinator to a device in a nonbeacon enabled network:

  In a nonbeacon-enabled network, the coordinator stores the data and waits for the device to request a data transfer. The device sends a request data command to the coordinator using slotted CSMA/CA. The coordinator responds to this request with an ACK. The coordinator then transfers data, and the device acknowledges having received the data, as shown in Figure 4.20.

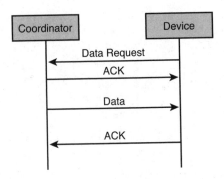

Figure 4.20   Coordinator to device data transfer in a nonbeacon-enabled network

4.6.2. Peer-to-Peer Data Transfer

A device attempting to transmit data to another device needs to synchronize itself with the other device before data transmission, and then the device uses unslotted CSMA/CA for transmitting data.

4.7 MAC FRAME FORMAT

The MAC layer defines four types of frame formats:

- **MAC data frame:** Used for transferring data
- **MAC beacon frame:** Generated by the coordinator for synchronization
- **MAC command frame:** Used by the MAC management entity
- **MAC acknowledge frame:** Acknowledges successful reception of the frame

4.7.1 General MAC Frame Format

Due to the similarities between MAC frame formats, it is beneficial to first describe the MAC frame format in the general case. As shown in Figure 4.21, the MAC layer adds its MAC header (MHR) and MAC footer (MFR) to those frame payloads it receives from the network layer.

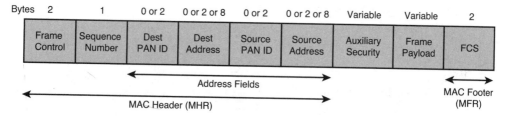

Figure 4.21   General MAC frame format

- **Frame Control field:** This 16-bit field is composed of the fields shown in Figure 4.22.

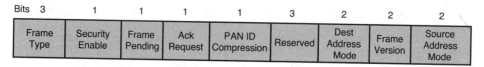

Figure 4.22   Frame Control field

- **Frame type:** Represented as 3 bits, it indicates the type of frame being transmitted:

  000   Beacon frame

  001   Data frame

  010   Acknowledge frame

  011   Command frame

  00–111   Reserved

- **Security enabled:** Indicates whether security is enabled. This field set to 1 means security has been applied to the frame at the MAC layer.

- **Frame pending:** If the value of this field is set to 1, there is more data being held for this device, and the device can request the data.

- **Acknowledge request:** When the value of this field is 1, the source is requesting an acknowledgment frame from the destination.

- **PAN ID compression:** This bit indicates if the destination device is located in the same network or a different network as the source device. When set to 1, the source PAN ID is the same as the destination PAN ID and not transmitted in the frame; otherwise, the destination PAN ID is different.

- **Frame version:** If set to 00, the frame conforms to the IEEE 802.15.4 2003 version. A value of 01 indicates adherence to the IEEE 802.15.4 2006 version.

- **Source and destination addresses mode:** This field defines the address size for the source and destination.

  PAN ID and address field for the destination are not present (indirect addressing).

  Reserved.

  Address field is 16 bits (short address).

  Address field is 64 bits (extended address).

- **Sequence number:** The sequence number of each frame.

- **Destination PAN ID:** If the frame is intended for a coordinator, this field represents the 16-bit coordinator ID. If the frame is intended for an end device, this field is discarded.

- **Source address and destination address:** These fields may either be the 16- or 64-bit source and destination addresses.

- **Auxiliary security:** This field represents the auxiliary security header, which contains the Security Control, Frame Counter, and Key Identifier fields.

- **Frame payload:** This field is variable depending on the type of the frame.

- **FCS (frame check sequence):** The FCS is used for error detection and is generated using the following cyclic redundancy check (CRC) polynomial:

$$\text{CRC-16: } X^{16} + X^{12} + X^5 + 1$$

4.7.2 Beacon Frame Format

The beacon frame is used for synchronization. It is generated by the coordinator on an interval basis to broadcast the presence of the coordinator to other devices. Figure 4.23 shows the beacon frame format.

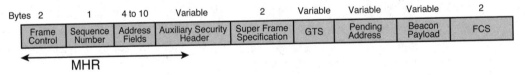

Figure 4.23 Beacon frame format

- **Frame control:** Same as the frame control depicted in Figure 4.22.
- **Sequence number:** The sequence number of the frame.
- **Address fields:** Contains the source PAN ID and address.
- **Superframe specification:** Figure 4.24 shows the superframe specification fields.
- **Beacon order (BO):** Used to calculate the BI; see section 4.5.1, "Superframe," for greater detail.

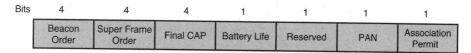

Figure 4.24   Superframe specification fields

- **Superframe order (SO):** Used to calculate the superframe duration (SD); again, see section 4.5.1 for greater detail.
- **Final CAP slot:** Represented as 4 bits, determines the duration of the CAP for the superframe.
- **PAN (true/false):** This bit, when set to 1, indicates the beacon frame is transmitted from the PAN coordinator.
- **Association permit (true/false):** When set to 1, indicates that the coordinator is accepting new devices. If set to 0, the coordinator will not accept any new devices.
- **Battery life extension (BLE) (true/false):** This field is set to 1 if the device can transmit to the coordinator during the CAP; otherwise, BLE should be set to 0.

4.7.3 MAC Data Frame Format

Figure 4.25 shows the MAC data frame format. For details regarding the Frame Control field and Address fields, see Figures 4.22 and 4.21, along with their corresponding descriptions.

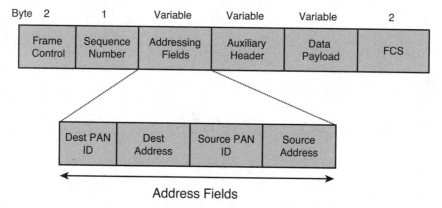

Figure 4.25   MAC data frame format

4.7.4 MAC Acknowledge Frame Format

Figure 4.26 shows the MAC acknowledge frame format.

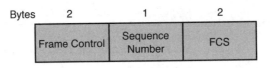

Figure 4.26 MAC ACK frame format

4.7.5 Command Frame Format

Table 4.2 lists the MAC layer command frames with their corresponding identifiers. An FDD can generate all the listed MAC commands, which are transmitted in the format shown in Figure 4.27.

Frame Control	Sequence Number	Address Field	Command Identifier	Command Payload	FCS

Figure 4.27 MAC command frame format

Table 4.2 MAC Layer Command Frames and Corresponding Identifier

Command Identifier	Command Name
01	Association request
02	Association response
03	Disassociation notification
04	Data request
05	PAN ID conflict notification
06	Orphan notification
07	Beacon request
08	Coordinator realignment
09	GTS request

4.8 ASSOCIATION REQUEST

When a coordinator or router is created, an end device can request association with it. If the end device is an orphaned device, it requests reassociation with the coordinator. To join the network, a device issues the association request command to the coordinator or router. Only an FFD can receive this association request. The coordinator responds to this request with the association response command. The following steps describe the association of a new device with a network:

1. The device issues an MLME-RESET.Request to set its MAC PIB default value to true.

2. Perform active or passive scan using MLME-SCAN.Request.

3. The result of the scan is used by the device to select the proper PAN.

Figure 4.28 shows the frame format for an association request.

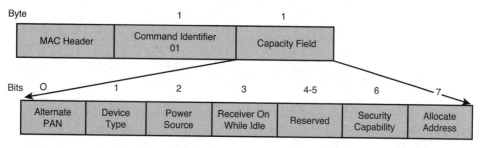

Figure 4.28   Association request frame format

- **Alternate PAN:** When this bit is set to 1, the device can become a coordinator.
- **Device type:** If a device requesting association is an FFD, this bit is set to 1; otherwise, this field is set to 0.
- **Power source:** This bit is set to 1 if the device's power is from main power; otherwise, it is set to 0.
- **Receiver on while idle:** This bit, when set to 1, will turn off the receiver during the idle period to save power; otherwise, the receiver will remain on.
- **Security:** When 1, this bit indicates the MAC layer should send and receive frames using the security protocol; if 0, no security protocol is used.
- **Allocation address:** Setting this bit to 1 indicates that the coordinator will assign a short address to the device. If set to 0, the device receives a short address of 0xFFFE in the association response; however, when the device receives the response from the coordinator, it will use its unique 64-bit address to communicate with coordinator.

4.8.1 Association Response

The coordinator responds to the device association request with the association response command. Figure 4.29 shows the association response frame format.

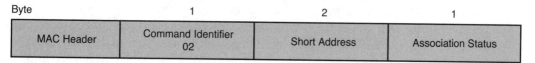

Figure 4.29   Association response frame format

- **Short address:** The address of the device that is requesting association; if the coordinator rejects the device association, this field is set to 0xFFFF.
- **Association status:**

    00 Association successful.

    01 AN or coordinator does not have capacity.

    02 Coordinator access was denied.

4.9 DISASSOCIATION NOTIFICATION COMMAND

Any device that is associated with the coordinator can send a disassociation command to the coordinator, or the coordinator may request a device for disassociation. Figure 4.30 shows the disassociation frame format.

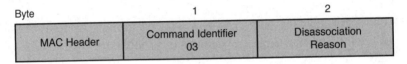

Figure 4.30   Disassociation frame format

- **Disassociation reason:**

    01 The coordinator is attempting to remove the device from the network.

    02 The device is attempting to leave the network.

4.10 ORPHAN NOTIFICATION

When a device loses its association with its parent, the device is an orphaned device. The loss of association can be caused by

- Parent is disabled.
- The device is moved out of range of its parent.

The device sends an orphan notification to the coordinator or router (parent) for reassociation. Figure 4.31 shows the reassociation frame format.

When the MAC layer of the coordinator or router receives the orphan notification command, it generates an MLME-ORPHAN.Indication, which it transmits to the network layer. The network layer determines whether the device was associated with the network. It then transmits an MLME-ORPHAN.Response to the MAC layer.

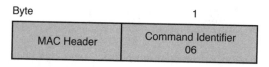

Figure 4.31   Reassociation frame format

4.11 BEACON REQUEST

This frame is transmitted by any device to determine whether any coordinator is operating within its personal operation space (POS). Figure 4.32 shows the frame format of a beacon request.

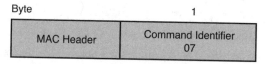

Figure 4.32   Beacon request frame format

4.12 COORDINATOR REALIGNMENT COMMAND

This command is issued for two reasons. It is sent as a response to the orphan notification command, where its recipient is the orphaned device, or as a response to an MLME-START.Request, where it is broadcasted to all devices in the network to indicate a change in the coordinator's attributes. Figure 4.33 shows the coordinator realignment command.

Byte	1	2	2	1	2	0 or 1
MAC Header	Command Identifier	PAN ID	Coordinator Short Address	Logical Channel	Short Address	Channel Page

Figure 4.33   Coordinator realignment command

- **Short address:** If this is a broadcast frame, the short address is set to 0xFFFF.
- **Channel page:** The channel page number that coordinator will use.

SUMMARY

- The MAC layer is located between the physical and network layers.
- It performs data transfer through the MAC common part sublayer (MCPS).
- It performs management services through the MAC layer management entity (MLME).
- The MAC layer uses the MCPS-DATA.Request, MCPS-DATA.Confirm, and MCPS-DATA.Indication primitives for data transfer.
- It performs device association and disassociation.
- Within the coordinator, it generates the beacon frame.
- It uses CSMA/CA as its device access method.
- It provides the coordinator with its ability to offer guaranteed time slots (GTS) to the devices in the network.
- The MAC layer contains the MAC Information Base (MIB) (database), which holds constant and variable MAC layer attributes.
- It generates the MLME-SCAN to initiate the scanning of channels by the physical layer.
- It sends the MLME-ASSOCIATE command to the coordinator for device association.
- The MLME-GET and MLME-SET commands are used to read from and write to the MIB.
- MLME-START command, generated by the network layer, causes the MAC layer to generate a new superframe.
- The MAC layer may use either slotted CSMA/CA (the beacon enable access method) or unslotted CSM/CA (the nonbeacon-enabled access method) for accessing the network.
- The MAC layer defines four frame format types: data, beacon, command, and acknowledgment.

REFERENCES

1. ZigBee Alliance, ZigBee Specification Document 053474r17, 2008

2. Daintree Network, Comparing ZigBee Specification versions

3. IEEE 802.15.4 Specification 2003

4. Farahani, S., ZigBee Wireless Network and transceivers, Newnes, 2008

5. Gutierrez, J., Gallaway, E., and Barrett, R., Low-Rate Wireless Personnel Area Networks, IEEE Press Publication, 2007

CHAPTER 5

NETWORK LAYER

INTRODUCTION

The network layer is located between the MAC layer and the application support sublayer (APS). It provides routing and establishes the ZigBee network topologies: star, mesh, and cluster tree. It starts a network, assigns node addresses, configures new devices, discovers other networks, and applies security. To fulfill these various responsibilities, the network layer offers data services through the network layer data entity (NLDE) and management services through network layer management entity (NLME). It also contains the Network Information Base (NIB) (database), which holds network layer attributes. Figure 5.1 shows the network layer interface as it relates to the MAC layer and the APS.

The NLDE provides data services to the APS for the transferring of data between two or more devices located within the same network. The NLME is responsible for generating the management primitives that manage the network. These primitives include those necessary for network discovery, network formation, joining a network, and leaving a network.

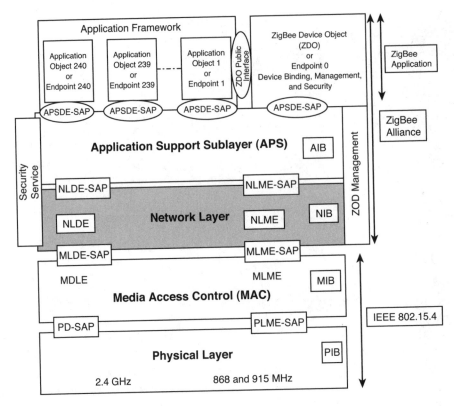

Figure 5.1   Network layer architecture

5.1 NETWORK LAYER DATA ENTITY (NLDE) SERVICES

The NLDE generates the network protocol data unit (NPDU) by accepting an application protocol data unit (APDU) from the APS and adding its header to the APDU. The NLDE then transmits the NPDU to the destination or next network hop. The network layer uses the following primitives for data service:

- NLDE-DATA.Request
- NLDE-DATA.Confirm
- NLDE-DATA.Indication

5.1.1 NLDE-DATA.Request

This command, generated by the APS, is sent to the network layer for the transfer of the protocol data unit (PDU). The network layer adds its header to the PDU, creating a network service data unit (NSDU). The network layer then transmits an NLDE-DATA.Confirm to the APS to indicate reception of the NSDU. Figure 5.2 shows the NLDE-DATA.Request primitive.

Dest Address Mode	Dest Address	NDSU Length	NDSU	NDSU Handle	Radius	Nonmember Radius	Discovery Route	Security Enable

Figure 5.2   NLDE-DATA.Request

- **Destination address mode:** Defines the type of destination address; 0x01 denotes group addressing, and 0x02 indicates unicast and broadcast addressing.
- **Destination address:** The destination address within the network represented by 16 bits.
- **NSDU length:** The number of bytes in the NSDU.
- **NSDU:** The actual data of the NSDU.
- **NSDU handle:** Identification for the NSDU; must be between 0x00 to 0xFF.
- **Radius:** The maximum number of hops the frame may travel.
- **Nonmember Radius:** The number of hops a multicast frame is allowed to travel; it may range from 0x00 to 0x07.
- **Discover route:**

  00 Suppress route discovery (use existing route)

  01 Enable route discovery (use current route unless there is no route, in which case perform route discovery)

- **Security enabled (true/false):** If the security enabled parameter in the NIB is set to 0, no security will be applied to the frame; otherwise, the network layer should apply security to the frame.

5.1.2 NLDE-DATA.Confirm

This is generated by the NLDE and sent to the APS to indicate the status of the NLDE.DATA.Request. Figure 5.3 shows this primitive.

Status	NSDU Handle	TX Time

Figure 5.3   NLDE-DATA.Confirm

- **Status:** Possible values include success, invalid request, and route error.

  When the network layer receives an NLDE-DATA.Request, it inspects the destination address to determine whether the destination node is associated with network. If it is not, the network layer sends an NLDE-DATA.Confirm primitive to the APS with a status of invalid request.

- **NSDU handle:** Corresponds to the NSDU handle in the NLDE-DATA.Request.
- **Tx time:** Indicates the time that the frame was transmitted. This field is used when the network time stamp is set to true in the NIB.

5.1.3 NLDE-DATA.Indication

This is sent by the network layer to the APS for the transfer of the NSDU. Figure 5.4 shows the frame format of the NLDE-DATA.Indication primitive.

Dest Address Mode	Dest Address	Source Address	NSDU Length	NSDU	Link Quality	Rx Time	Security

Figure 5.4   NLDE-DATA.indication

- **Destination address mode:** A value of 0x01 indicates multicast addressing, and 0x02 indicates unicast and broadcast addressing.
- **Source address:** Address of the source device.
- **Link quality:** Indicates the signal quality of receiving frame.
- **Rx time:** Indicates the time that the frame was received.
- **Security (true or false):** True indicates that network security is applied to the incoming frames. False indicates that network security is not applied to the incoming frame.

5.2 NETWORK INFORMATION BASE (NIB)

The NIB holds the network layer attributes. These attributes may either be constant or variable.

5.2.1 NIB Constant Attributes

The following are some of the constant attributes stored in the NIB:

- **Network coordinator ability:** A value of 0x00 indicates that the device is incapable of becoming a coordinator. A value of 0x01 means it can become a coordinator.
- **Network discovery retries:** Set to 0x03, it defines the number of allowable retries for route discovery.
- **Network default security**: Indicates the default security level (set to 5).

- **Minimum header:** The minimum number of bytes the network layer adds to the NSDU.

- **Network layer protocol version:** The current version is 0x02.

- **Network minimum header overhead:** Set to 0x08, it is the minimum number of bytes contained within the NSDU.

- **Network route discovery time:** Set to 2,710 milliseconds, it defines the maximum duration of a route discovery operation.

- **MAC frame header:** Indicates the number of bytes of MAC header, and its value is 0x0B.

- **Route request retries**: Number of times a route request can be retried, and its value is 0x02.

5.2.2 NIB Variable Attributes

The following attributes are used to manage the network layer. They can be read from using the NLME-GET.Request command and can be written to using the NLME-SET.Request command:

- **Network sequence number:** The sequence number of the outgoing frame. It is a random number from 0x00 to 0xFF.

- **Network broadcast retries:** Ranges between 0x00 and 0x05; its default value is 0x03.

- **Network passive acknowledge time:** The length of time the transmitter should wait for an acknowledge frame. It is between 0x0000 and 0x2710 in milliseconds

- **Network maximum router:** Defines the maximum number of routers in the network. Its value can be between 0x00 and 0xFF; the default is 0x05.

- **Network tree address allocation (true/false):** A value of true indicates the network layer assigns addresses to the devices; a value of false means a higher layer assigns the addresses.

- **Network maximum children:** The maximum number of children allowed in the network.

- **Network maximum depth:** The maximum depth of the network using a tree topology. By default, it is set to 0xFF. This attribute does not apply to the ZigBee PRO stack.

- **Network neighbor table:** This table contains the following information:
64-bit IEEE address.

Network address.

Device type (coordinator, router, and end device).

Receiver on while idle (true or false).

Relationship. (0x00 neighbor is parent, 0x01 neighbor is a child, 0x02 neighbor is a sibling, 0x04 neighbor is a previous child, and 0x05 neighbor is an unauthenticated child.)

Link quality indication.

Outgoing cost. (Indicates the cost of the neighbor link. If the symmetry link is set to true, this field is mandatory.)

Dept (used in a tree topology).

Logical channel operation of the neighbor.

- **Network report constant cost:** If set to 00, the network layer will calculate the cost from its link to neighbor nodes. If set to 01, use the constant value. For ZigBee PRO, this is set to 0.
- **Network symmetric link (true/false):** True means the route is a symmetric link. It is created during route discovery for forward and backward routing. False means the route is asymmetric and only the forward link is created during route discovery.
- **Network capability information:** This field is 8 bits (B_7 B_6 B_5 B_4 B_3 B_2 B_1 B_0)

The following describes the function of each bit in the network capability information:

B_0 **(alternate PAN):** Set to 0.

B_1 **(device type):** 1 means it is a ZigBee router joining the network; 0 means it is an end device joining the network.

B_2 **(power source):** 1 means the device uses main power; 0 means another type of power source.

B_3 **(receiver on when device idle):** 1 means the receiver is on when the transceiver is idle; 0 means the receiver is off when the transceiver idle.

B_5 B_4**:** Reserved.

B_6 **(security):** 0 means the device uses standard security; 1 means the device uses high security.

B_7 **(address allocation):** 1 means the parent of the device allocates an address to the device.

- **Network routing table:** Routing table entries.
- **Network address allocation:** Defines the method that the network uses to allocate addresses to the devices. 0x00 means it uses a distribution scheme (Cskip) and 0x02 means it uses the Stochastic (random) address assignment. The stochastic method is used by ZigBee PRO.
- **Network use tree routing (false/true):** True means use tree routing. False means do not use tree routing. For ZigBee PRO, this field is set to false.
- **Network manager address**: Address of the manager node.
- **Network short address:** The 16-bit address of the device.

- **Network stack profile:** For ZigBee PRO, it is set to 0x02
- **Network maximum source route:** The number of hops a packet may travel to reach the destination. It is used in the source routing algorithm.
- **Network broadcast transaction:** The ZigBee coordinator, router, or end device that has its receiver on while idle should keep a record of all broadcast transactions (received from the neighbors or transmitted). This information is called a broadcast transaction record (BTR) and contains the following information: source address of the broadcaster message (16 bits), sequence number, and expiration time.
- **Network group ID:** 16-bit group membership number.
- **Network uses multicast (true/false):** True means the multicast message is generated at the network layer; false means the multicast message is generated at the APS.
- **Network is concentrator (true/false):** True means the device is a concentrator. False means the device is not a concentrator.
- **Network concentrator radius:** Number of hops from the device to the concentrator.
- **Network address map:** Holds the 64-bit IEEE address and the 16-bit network address of the device.
- **Network PAN ID:** 16-bit network ID
- **Network transaction persistent time:** Defines the maximum time that a transaction can be stored in the coordinator. It is indicated by the beacon frame.

5.3 NETWORK LAYER MANAGEMENT ENTITY (NLME)

The NLME performs the following functions.

- **Configuration of new devices:** Configures a new device so as to be the coordinator or simply as a device joining to the network.
- **Starting a network:** Only performed by the network's coordinator.
- **Addressing:** The coordinator and routers can assign addresses to each end device joining the network.
- **Neighbor Discovery:** Discovers one-hop neighbors, recording their addresses and capabilities.
- **Route Discovery:** Finds the most efficient route over which to transfer messages to a destination.
- **Security:** Applies the security protocol to outgoing and incoming frames.
- **Joining and leaving a network:** A device has the capability to join or leave a network.

The NLME provides the following primitives for managing the network layer.

NLME-NETWORK-DISCOVERY

NLME-NETWORK-FORMATION

NLME-PERMIT-JOINING

NLME-START-ROUTER

NLME-DIRECT-JOIN

NLME-LEAVE

NLME-NETWORK-MANAGEMENT

NLME-PERMIT-JOINING

NLME-RESET

NLME-SET

NLME-START-ROUTER

NLME-SYNC

NLME-ROUTE-DISCOVERY

5.3.1 Network Discovery

When the ZigBee upper layer sends a discovery request command (NLME_NETWORK-DISCOVERY.Request) to the network layer, the network layer searches for a network within the personal operation space (POS). When the scanning is completed, the NLME sends a network discovery confirm to the upper layer. Figure 5.5 shows the NLME-NET-WORK_DISCOVERY.Request primitive.

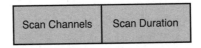

Figure 5.5   Network discovery request

- **Scan channels:** This 32-bit field, represented by $B_{31}B_{30} \ldots B_{27}B_{26} \ldots B_0$, indicates which channels must be scanned. When B_i is set to 1, the ith channel must be scanned for i less than 27.
- **Scan duration value:** Ranging from 0x00 to 0x0E, this field, along with the base superframe duration, is used to calculate how long each channel may be scanned. Equation 5.1 is used to calculate this scan duration, where the base superframe duration is defined as the number of symbols in a single slot (60).

Scanning time = (2 * scan duration value + 1) * base superframe duration   (5.1)

Figure 5.6 shows the network discovery process, which is as follows:

1. The higher layer requests network discovery by sending the NETWORK-DISCOVERY. Request to the NLME, which includes a list of channels and the scan duration of each channel.

2. The NLME sends an MLME-SCAN request to the MLME for an active scan. An active scan is used to locate a coordinator in its POS.

3. The MLME sends an MLME beacon request command to the physical layer for transmission of this command.

4. The physical later transmits the beacon request command and turns on its receiver to receive a beacon.

5. If a beacon frame is received by the MLME with no payload, the MLME sends an MLME-BEACON-NOTIFY to the network layer.

6. Step 2 through 5 are repeated for each channel. The complete list of networks within the POS is transmitted to the upper layer with an NLME-NETWORK-DISCOVERY.Confirm. Figure 5.7 shows the NLME-NETWORK-DISCOVERY.Confirm primitive.

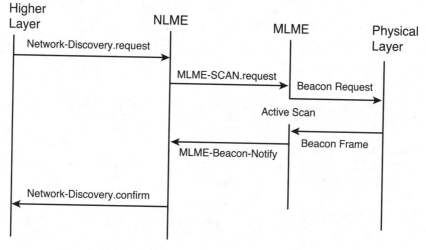

Figure 5.6   Network discovery process

Status	Network Count	Network Description

Figure 5.7   NLME-NETWORK-DISCOVERY.Confirm

- **Status:** This is the status of the NLME-SCAN.Confirm.
- **Network count:** This field contains the number of networks found by the network discovery operation.
- **Network descriptor:** The network descriptor contains information about the networks that were found, including the following fields:

  Extended PAN ID: The 64-bit PAN identifier.

  Logical channel: The current logical channel used by the network.

  Stack profile: The ZigBee stack profile identifier, which is used for network discovery.

  ZigBee version: The current version of ZigBee; currently, it is 2.0.

  Beacon order (BO): The BO is used to determine how often the MAC layer transmits the beacon frame.

  Superframe order (SO): The SO is used in beacon-oriented networks; it defines the length of the active portion of superframe.

  Permit joining: Indicates whether the coordinator of the discovered network allows new devices to join its network.

  End device capacity (true/false): True means the device is able to accept the NLME-JOIN.Request command; otherwise, it does not.

  Route capacity (true/false): True means the device has routing capability. False means it does not.

5.4 NETWORK FORMATION

To create a network, the higher layer of the ZigBee coordinator sends an NLME-NETWORK-FORMATION.Request to the network layer. Figure 5.8 shows the NLME-NETWORK-FORMATION.Request primitive.

Scan Channels	SCAN Duration	Superframe Order	PANID	Battery Life Extension

Figure 5.8 NLME-NETWORK-FORMATION.Request

- **Scan channels:** Indicates which channels need to be scanned.
- **Scan duration:** Determines the scan duration for each channel.
- **Beacon order:** Indicates the frequency at which the coordinator transmits the beacon frame.

- **Superframe order:** Defines the active portion of the superframe, which is determined by a higher layer.

- **Battery life extension (true/false):** This used by those devices that generate a beacon frame, namely coordinators. True indicates that the coordinator's receiver is disabled during the full backoff period; false indicates that the receiver is enabled during the contention access period (CAP).

NLME-NETWORK-FORMATION.Confirm: This is sent by the NLME to the APS in response to a network formation request. It contains the status of the request: invalid request, startup failure, or success.

5.4.1 Network Formation Process

If the device requesting network formation is not a coordinator, the NLME issues an NLME-NETWORK-FORMATION.Confirm containing a status of invalid request. However, if the device is a coordinator, the following process is used to form the network:

1. The coordinator requests network formation by sending an NLME-NETWORK-FORMATION.Request to the network layer.

2. The network layer sends an MLME-SCAN.Request to the MAC layer indicating a request to perform energy detection on selected channels. The MAC layer performs the energy detection for each channel and sends the results to the network layer via an MLME-SCAN.Confirm.

3. The network layer accepts those channels that have acceptable energy levels.

4. By sending another MLME-SCAN.Request to the MAC layer, the network layer requests that the MAC layer perform an active scan on those channels found in step 3.

5. The MAC layer scans the channels and submits the list of available channels by sending an MLME-SCAN.Confirm to the network layer.

6. The network layer selects the best channel and assigns a 16-bit network address (PAN ID); it then sends an MLME-SET.Request to update the MIB.

7. After the MAC layer updates its information database (MIB), it responds to the network layer with an MLME-SET.Confirm.

8. The network layer sends an MLME-START.Request.

9. The MAC layer responds to the network layer with an MLME-START.Confirm, causing the network layer to send an NLME-NETWORK-FORMATION.Confirm to the APS. The NLME-NETWORK-FORMATION.Confirm indicates the status of the network formation.

5.5 JOINING A NETWORK

A device can join a network in three different ways:

- By requesting to join a network through association.
- By joining a network directly.
- An orphaned device may request to rejoin a network. When a device loses its connection from its parent, it becomes an orphaned device. This is generally the result of the parent becoming disabled or the device moving out of range of the parent. If a device is orphaned, it will do an orphan scan by broadcasting an "orphan notification" command frame in the hopes of finding its parent. If the parent gets the notification, it will inform the device that it is still there and the orphan can rejoin that parent.

In the direct joining method, the parent of the device is preprogrammed with the 64-bit address of the child device which allows the parent to send an NLME-DIRECT-JOIN.Request to the child to join the network. The following steps describe the process for a device joining a network:

1. The application layer of the device issues an NMLE-DISCOVERY.Request to the network layer.

2. When application layer receives an NLME-DISCOVERY.Confirm, it selects a network.

3. After the network is selected, the application layer of the device issues an NLME-JOIN. Request to its network layer. Figure 5.9 shows the NLME-JOIN.Request primitive.

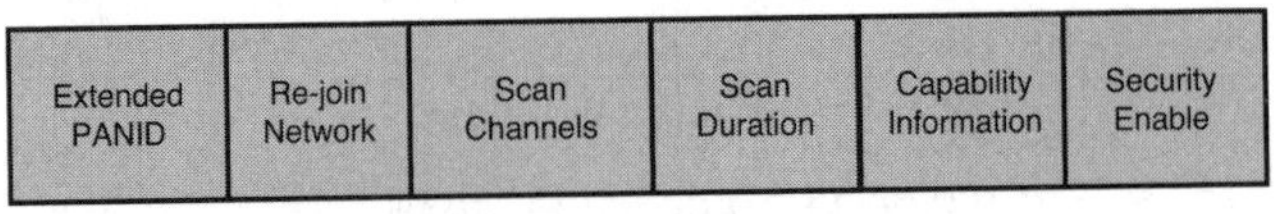

Extended PANID	Re-join Network	Scan Channels	Scan Duration	Capability Information	Security Enable

Figure 5.9   NLME-JOIN.Request

- **Extended PAN ID (64 bits):** The network ID of the network the device wants to join.
- **Join as router:** When true, the device joins the network as a router.
- **Rejoin network:** Ranges from 0x00 to 0x03. It indicates the type of joining process:

  0x00   The device requests joining through association.

  0x01   The device requests joining directly or as an orphan.

  0x02   The device requests joining the network using the network layer rejoin process. This is used by an end device that lost its connection from the network and its parent.

  0x03   The device requests switching the operational channel.

- **Scan channels:** Indicates which channels must be scanned.
- **Scan duration:** Indicates how long a channel must be scanned.
- **Power source:** A value of 0x01 indicates the use of a main power source, whereas 0x00 indicates the use of some other power source.
- **Rx on when idle:** 0x00 means the receiver is enabled when the device is idle; 0x01 means the receiver is disabled while the device is idle.
- **MAC security:** When 0x01, MAC security is enabled; otherwise, it is disabled.

Before a device may join the network, the coordinator or router must advertise permission to accept new devices into the network. A higher layer of the ZigBee coordinator or router sets the MAC association permission flag in the MIB to either true or false by transmitting an NLME-PERMIT-JOINING.Request to the NLME. The NLME then sends an MLME-SET.Request to the MLME for setting the MIB value. The NLME-PERMIT-JOINING.Request contains the permit duration value (0x00–0xFF). A value of 0x00 indicates that permission to join the network is disabled, and a value of 0xFF indicates that it is enabled. Figure 5.10 shows the NLME permit joining process.

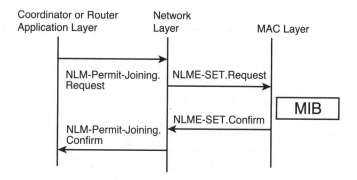

Figure 5.10   NLME permit joining process

In Figure 5.10, the NLME-PERMIT-JOINING.Confirm indicates the status of the operation.

5.5.1 NLME–LEAVE.Request

This is sent from the APS of a device to the network layer to indicate that the device wants to leave the network. It may also be sent by the coordinator or router to request that a device leave the network.

5.5.2 NLME–LEAVE.Indication

This primitive is used by the coordinator or a router to indicate to an end device that it has been removed from the network. This primitive contains the device address (64-bit IEEE address) and rejoin (true/false). The rejoin value indicates that the device may join the network again or not.

5.5.3 NLME–RESET.Request

This is issued by the APS to the network layer for the purpose of resetting the NIB parameters to their default values, clearing the routing tables and, by generating an MLME-RESET.Request, resetting the MIB parameters to their default values.

5.5.4 NLME–START–ROUTER.Request

This primitive, generated by the APS, is used to set a device as a router, allowing it to route data and perform route discovery.

5.5.5 Managing the Network Information Base

The NLME-GET.Request is used to read NIB attributes, and the NLME-SET.Request is used to change NIB attributes. See section 5.2, "Network Information Base (NIB)," for greater detail.

5.6 NETWORK LAYER FRAME FORMAT

The network frame is transmitted to MAC layer for transmission. There are two types of network frames: data and command frames. The network layer frame format consists of the network header and a payload, as shown in Figure 5.11

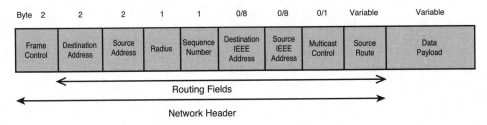

Byte 2	2	2	1	1	0/8	0/8	0/1	Variable	Variable
Frame Control	Destination Address	Source Address	Radius	Sequence Number	Destination IEEE Address	Source IEEE Address	Multicast Control	Source Route	Data Payload

Routing Fields

Network Header

Figure 5.11   Network layer frame format

- **Frame control**: Figure 5.12 shows the Frame Control field in greater detail.

Bit 2	4	2	1	1	1	1	1	3
Frame Type	Protocol Version	Discover Rout	Multicast Flag	Security	Source Route	Destination IEEE Address	Source IEEE Address	Reserved

Figure 5.12   Frame Control field

- **Frame type:**

  00   Data frame

  01   Network command frame

  10 and 11   Reserved

- **Protocol version:** This defines the ZigBee protocol version, and its value is in the NIB.

- **Discover route:** The network layer offers route discovery to find the best route for transmitting a message to a destination. There are three possible values for route discovery:

  Suppress route discovery (00): Use the current route.

  Enable route discovery (01): The message is routed through the current route; if there is no route, the router should start a new route discovery.

  Force router discovery (11): The router starts route discovery even if the router already has a route.

- **Multicast:** If true, the frame is a multicast frame; otherwise, it is a unicast or broadcast frame.

  The broadcast frame can have following addresses:

  Address 0xFFFF All devices in the network will accept the frame.

  Address 0xFFFD Devices with their receivers on while they are idle accept the frame.

  Address 0xFFFC Coordinator and routers accept the frame.

  Address 0xFFFB Low-power routers accept the frame.

- **Source route:** If true, the network header contains the route to the destination.

- **Destination IEEE address:** If true, the destination address is an IEEE address.

- **Source IEEE address:** If true, the source address is an IEEE address.

- **Security field:** If this bit set to 1, the network layer will apply security to the outgoing frame.

- **Destination address:** Represents the 16-bit network address of the destination.

- **Source address:** This represents the network address of the source.

- **Radius:** This defines the maximum number of routers a message may travel to reach its destination (same as Time To Live [TTL] in IP packets).

- **Sequence number:** This value is incremented each time a frame is transmitted.

- **Destination IEEE address:** The actual destination address; if the control field for the destination IEEE address is 1, this field contains the 64-bit IEEE address.

- **Source IEEE address:** The actual source address; if the control field for the source IEEE address is 1, this field contains the 64-bit IEEE address.

- **Multicast Control field:** If the Multicast Flag field shown in Figure 5.12 is set to true, the field depicted in Figure 5.11 is present; otherwise, it is not present. The Multicast Control field indicates whether the destination devices belong to a member group.

5.7 NEIGHBOR TABLE

Each device contains information about those devices located within its transmission range. This information is held in a table called the neighbor table. Each entry in the table holds the following information:

- **PAN ID:**
  64-bit IEEE address of the device
  16-bit network address
- **Device type:**
  00   ZigBee coordinator
  01   Router
  02   End device
- **While idle (true/false):** Indicates if the neighbor device receiver is on while the device is idle. True means the receiver of the device is on; false means the receiver is off.
- **Relationship:** Defines the relationship between the neighbor and the device, as follows:
  0x00   Neighbor is a parent.
  0x01   Neighbor is a child.
  0x02   Neighbor is a sibling.
  0x04   Neighbor is a previous child.
  0x05   Neighbor is an unauthenticated child.
- **Transmission failure**: Indicates previous transmissions failed or succeeded.
- **Link quality indicator (LQI):** Indicates the link quality of the neighbor.
- **Outgoing cost:** Indicates the cost of link. For symmetric links, this field is mandatory.
- **Depth of device:** Number of hops from the device to the coordinator.
- **Permit joining (true/false):** A value of true means the neighbor device accepts new devices joining the network; false means the neighbor device does not accept new devices.
- **Logical channel:** Channel at which the neighbor device is operating.

5.8 NETWORK COMMAND FRAME FORMAT

Figure 5.13 shows the network layer command frame format where the Frame Control field is similar to Figure 5.12 and the Routing field is the same as the one depicted in Figure 5.11.

Frame Control	Routing Field	Network Command Identifier	Network Command Payload

Figure 5.13   Network command frame format

- **Network command identifier:**

  01   Route request

  02   Route reply

  03   Route error

  04   Leave (leaving network)

  05   Route record

  06   Rejoin request

  The above commands are transmitted to the MAC layer by the MCSP-DATA.Request.

- **Leaving a network:** A device can leave a network in two ways, as follows:

  - *The device can request to leave a network:*

    A device can inform the ZigBee router or coordinator that is leaving the network by sending an NLME-LEAVE.Request command. The following steps describe this leave process:

    1. The APS sends an NLME-LEAVE.Request to the network layer.

    2. The network layer sends this command to the MAC layer using an MCPS-DATA.Request with the destination address set to 0xFFFF (broadcast).

    3. The MAC layer transmits the PD-DATA.Request to the physical layer for a transmission of the leave command and transmits the MCPS-DATA.Confirm to the network layer.

    4. The network layer sends an NLME-LEAVE.Confirm to the APS.

  - *The parent (router or coordinator) of the device requests that the device (child) leave the network:*

  The following steps describe this leave process:

    1. The APS sends an NLME-LEAVE.Request command to the network layer with the 64-bit IEEE address of the child. The network layer checks the relationship value of its neighbor table. If this is not 0x05, go to Step 2. If the relationship value of its neighbor table is 0x05, the device is unauthenticated and the router or coordinator cannot send the leave command to the device.

    2. The network layer sends an MCPS-DATA.Request with the 16-bit network address of the child.

SUMMARY

- The network layer is located between the media access control (MAC) layer and application support sublayer (APS).

- It provides routing, network formation, address assignment, new device configuration, and network discovery.

- It offers data services through the network layer data entity (NLDE) and management services through the network layer management entity (NLME).

- The Network Information Base (NIB) (database) holds network layer attributes.

- The available network layer data services are the NLDE-DATA.Request, NLDE-DATA.Confirm, and NLDE-DATA.Indication.

- The NLDE-DATA.Request is generated by the APS and sent to the network layer for the transmission of the application protocol data unit.

- The NLDE-DATA.Confirm is generated by the network layer and sent to the APS in response to NLDE-DATA.Request and to indicate the request status.

- The NLDE-DATA.Indication is generated by the network layer and sent to the APS for the transmission of the Network Protocol Data Unit (NPDU).

- The NIB holds constant and variable attributes. Variable attributes are the only ones that may be modified.

- Some NIB constant attributes are network coordinator ability, network discovery retries, network maximum depth, and network protocol version.

- Some NIB variable attributes are network broadcast retries, network maximum routers, network tree address allocation, network maximum children, network routing table, and network available addresses.

- NLME-NETWORK-DISCOVERY, NLME-NETWORK-FORMATION, NLME-PERMIT-JOINING, NLME-START-ROUTER, NLME-DIRECT-JOIN, and NLME-LEAVE are some of the NLME primitives.

- Network formation is performed by a coordinator's network layer upon receipt of an NLME-NETWORK-FORMATION primitive sent from its application layer.

- There are three possible ways by which a device may join a network: joining a network through association, joining a network directly, and an orphan device requesting to rejoin a network.

- A ZigBee application can transfer periodic data, intermittent data, and repetitive data.

- The network layer assigns a 16-bit address to each device.

- The network layer assigns its 16-bit PAN ID.

REFERENCES

1. ZigBee Alliance, ZigBee Specification Document 053474r17, 2008
2. Daintree Network, Comparing ZigBee Specification versions
3. IEEE 802.15.4 Specification 2003
4. Farahani, S., ZigBee Wireless Network and Transceivers, Newnes, 2008
5. Gutierrez, J., Gallaway, E., and Barrett, R. Low-Rate Wireless Personnel Area Networks, IEEE Press Publication, 2007

CHAPTER 6

ZIGBEE APPLICATION SUPPORT SUBLAYER (APS)

INTRODUCTION

The application support sublayer (APS) provides services to the application layer and the network layer through the application support data entity (APSDE) and application support management entity (APSME). Figure 6.1 shows the APS.

6.1 APPLICATION SUPPORT DATA ENTITY (APSDE)

The APSDE accepts the application protocol data unit (APDU) from the application layer, adds its header information to the APDU, and transfers the resulting frame to the network layer. The APSDE uses the following primitives for data transfer:

- APSDE-DATA.Request
- APSDE-DATA.Confirm
- APSDE-DATA.Indication

APSDE-DATA.Request: This is generated by the application object and sent to the APS for the purpose of transferring data. Figure 6.2 shows the APSDE-DATA.Request.

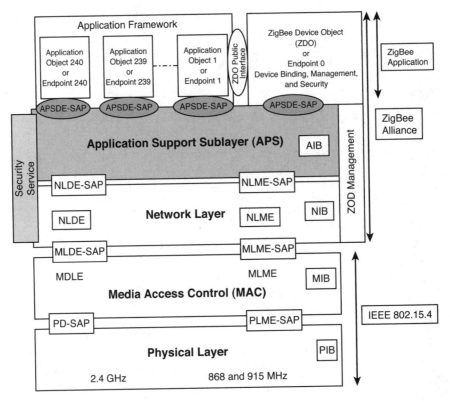

Figure 6.1 Application support sublayer

Dest Address Mode	Dest Address	Dest Endpoint	Profile ID	Cluster ID	Source Endpoint	asdu Length	asdu	Tx Option	Radius Counter

Figure 6.2 APSDE-DATA.Request

- **Destination address mode (8 bits):** This field can range from 0x00 to 0xFF. The following values are defined:

  0x00 The destination address and endpoint fields are not present in the frame but can be found in the supporting device binding table. The binding table of the device is searched using the cluster ID and source endpoint. If an entry or entries are found in the binding table, the APSDE uses the information to make the frame format for the APDU and transmits the resulting frame to the destination. In those instances when there is more than one entry, the application support sublayer transmits the APDU to each destination. The source device inspects its binding table to determine whether the destination address is a unicast or group of addresses.

0x01   The destination address is a 16-bit address representing a group of destination devices, and the destination endpoint address is not present.

0x02   The destination address is a 16-bit address, and the destination endpoint is part of the primitive.

0x03   The 64-bit destination address and destination endpoint are present. The 64-bit address is mapped to 16-bit address, which is located in the Network Information Base (NIB).

0x04 through 0xFF are reserved.

- **Destination address:** The address of the destination device; the size of the address is defined by the destination address mode.

- **Destination endpoint:** When the destination address mode is 0x02 or 0x03, this field, ranging from 0x00 to 0xFF, holds the address of the destination endpoint (application object); otherwise, it is empty.

- **Profile ID:** The application identification that is used by two endpoints to communicate with each other; the endpoint profile is a 16-bit number.

- **Cluster ID:** The set of attributes in an application profile is called a cluster. For example, the on and off positions of a switch are attributes, the set of which is referred to as a cluster.

- **Source endpoint:** Defines the source application object.

- **Application support data unit (ASDU) length:** The number of bytes in the ASDU field. If the number of bytes in the ASDU is more than the maximum number of bytes for each APS, the payload is fragmented.

- **ASDU:** Application support data unit.

- **Tx Option:** Security transmission option, which is represented by $0000\ B_3\ B_2\ B_1\ B_0$, where

  $B_0 = 1$ Security is enabled during transmission.

  $B_1 = 1$ The network key is used for encryption.

  $B_2 = 1$ An ACK transmission is requested.

  $B_3 = 1$ Fragmentation is permitted. When the application data unit is larger than the maximum size permitted by a single frame, the data is fragmented and transmitted as several frames.

- **Radius counter:** Defines how many hops from the source a packet can travel.

APSDE-DATA.Confirm: This is generated by the APS in response to the APSDE-DATA.Request. This primitive indicates the status of the request (for example, success or failure). Figure 6.3 shows the APSDE-DATA.Confirm primitive. For descriptions of the destination address mode, destination address, destination endpoint, and source endpoint, see the APSDE-DATA.Request frame format field descriptions.

Dest Address Mode	Dest Address	Dest Endpoint	Source Endpoint	Status

Figure 6.3 APSDE-DATA.Confirm primitive

- **Status:** Shows the status of APSDE-DATA.Confirm.

APSDE-DATA.Indication: This primitive is generated by the APS to indicate the transfer of a packet to an endpoint. Figure 6.4 shows the APSDE-DATA.Indication frame format.

Dest Address Mode	Dest Address	Dest Endpoint	Source Address Mode	Source Address	Source Endpoint	Profile ID	Cluster ID	ASDU Length	ASDU	Status	Security Status	Link Quality	Rx Time

Figure 6.4 APSE-DATA.Indication primitive

- **Status:** Indicates the status of an incoming packet.
- **Security status:** Indicates the security status of the ASDU received by the endpoint. Security statuses include ASDU received without any security, ASDU secured by network key, or ASDU secured by link key.
- **Link quality:** Indicates the quality of the link for incoming packets.
- **Rx Time:** Indicates the time that packet was transmitted (based on the source clock).

6.2 APPLICATION SUPPORT SUBLAYER MANAGEMENT ENTITY (ASME)

The ASME provides management of the Application Information Base (AIB), binding and unbinding of end devices, and the managing of group addressing. The ASME uses the following management primitives:

APSME.BIND.Request	APSME-UNBIND.Confirm
APSME-BIND.Confirm	APSME-ADD-GROUP.Request
APSME-GET.Request	APSME-ADD-GROUP.Confirm
APSME-GET.Response	APSME-REMOVE-GROUP.Request
APSME-SET.Request	APSME-REMOVE-GROUP.Confirm
APSME-SET.Response	APSME-REMOVE-ALL-GROUP.Request
APSME-UNBIND.Request	APSME-REMOVE-ALL-GROUP.Confirm

- **APSME-BIND.Request:** This management entity is generated by a ZigBee device object (ZDO) to request the binding of two endpoints or the addition of a binding record to the binding table; binding is a logical connection between two devices. Figure 6.5 shows the binding request primitive.

Source Address	Source Endpoint	Cluster ID	Dest Address Mode	Dest Address	Dest Endpoint

Figure 6.5   APSME-BIND.Request

As shown in Figure 6.5, both the source and the destination endpoints share a single cluster ID known as the link cluster.

Source address: The source address is a 64-bit IEEE address.

- **APSME-BIND.Confirm:** This command is generated by the APSME in response to a bind request. It shows the status of the binding (for example, success, binding table is full, or illegal request).

- **APSME-UNBIND.Request:** This is generated by a ZDO and issued to the APS to request the unbinding of two endpoints or to remove an entry from its local binding table.

- **APSME-UNBIND.Confirm:** Contains the status of the APSME-UNBIND.Request

- **APSME-GET.Request:** This is used by the APS to read the contents of the APS Information Base (AIB). As the name implies, the AIB contains attributes for the APS.

- **APSME-ADD-GROUP.Request:** This is generated by the application layer and sent to the APS for the purpose of adding an endpoint to a group. The APSME is able to manage many endpoints by granting them membership to a group. Specifically, this primitive is used to assign a group number to an endpoint. It contains the 16-bit group address and the address of the endpoint. This enables a ZigBee device to send a single packet to multiple endpoints. For example, an on and off switch controls multiple lights, and the lights belong to a group, which allows the switch to transmit a single packet and change the status of the lights.

- **APSME-ADD-GROUP.Confirm:** This is sent to the application layer in response to APSME-ADD-GROUP request. It contains the status of adding the endpoint to a group.

- **APSME-REMOVE-GROUP.Request:** Removes an endpoint's group membership.

- **APSME-REMOVE-ALL-GROUP.Request:** Removes all endpoints from a group membership.

6.3 APPLICATION SUPPORT SUBLAYER INFORMATION BASE (AIB)

The AIB contains both constant and variable attributes. The variable attributes can be modified using the ASME-SET.Request, and both attribute types can be retrieved using the ASME-GET.Request.

6.3.1 Application Support Sublayer Constant Attributes

The following are APS constant attributes:

- **APS maximum descriptor size:** Set to 64 bytes.
- **APS ACK wait time:** Specifies the length of time a device should wait for the ACK frame after transmitting a frame.
- **Maximum window size:** This is used for fragmented packets and defines size of buffer.
- **Interframe delay:** The delay between fragmented frames
- **Maximum number of retransmission attempts allowed:** Set to 3.
- **Maximum waiting time for ACK frame:** 0.1 seconds.
- **Minimum header overhead:** The minimum number of bytes for the APS header.

6.3.2 Application Support Sublayer Variable Attributes

The following are the APS variable attributes; these attributes can be read and written to using APSME-GET.Request and APSME-SET.Request.

- **APS application binding table:** The binding table of the device.
- **APS designated coordinator:** True or false. When true, the device should become coordinator on power up; when false, the device will not be the coordinator.
- **APS channel mask:** Defines the channels used by the network.
- **APS extended PAN ID:** The PAN ID is the 64-bit network ID, which has a default value of 0.
- **APS application group table:** The group table for the end device.
- **APS insecure join:** True or false. When true, a device can join a network in secure mode, and when false the device can join a network in unsecure mode.
- **APS interframe delay:** Defines the standard delay between two fragmented frames; it is set by the stack profile.
- **APS last channel energy:** Indicates the energy measurement of the previous channel.
- **APS channel timer:** Indicates a timer that is used for the scheduling of channel changes.

6.4 PERSISTENT DATA

The APS must hold the following information even after losing power. That is, the data must be persistent:

- Whether the device has a binding table
- Whether the device supports a designated coordinator
- Channel mask
- Extended PAN ID
- Group table if supported by device
- Binding table cache
- Node descriptor
- Power descriptor
- Simple descriptor

6.5 APPLICATION SUPPORT SUBLAYER FRAME FORMAT

The APS includes the data, command, and acknowledge frame formats. The Frame Type field within the Frame Control field defines the type of frame.

Figure 6.6 shows the APS general frame format, which consists of the APS header and the APS payload.

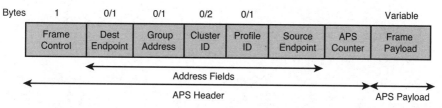

Figure 6.6   APS frame format

- **Frame control:** Figure 6.7 shows the frame control format contained within the general APS frame format.

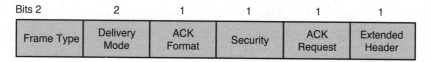

Figure 6.7   Frame control

- **Frame type (2 bits):** 00 data, 01 command, 10 ACK, and 11 reserved.
- **Delivery mode (2 bits):** 00 unicast, 01 indirect addressing, 10 broadcast, and 11 group addressing. For delivery mode 00 and 10, the frame should contain the source and destination endpoint addresses.

- **ACK format:** This field defines the format of acknowledge frame; if set to 1, the ACK frame should contain the cluster ID, profile ID, and source endpoint address.
- **Security (1 bit):** Setting this bit to 1 enables security on the APS frame.
- **ACK request (1 bit):** When true, the source requests an ACK from the destination.
- **Extended header:** If this bit set to 1, the frame contains an extended header. Figure 6.8 shows the extended header.

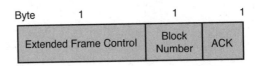

Figure 6.8 Extended header

- **Extended frame control:** 8 bits of the form XXRRRRRR, where Rs are reserved bits and

  XX = 00   The transmission did not fragment.

  XX = 01   The frame is the first one of a fragmented frame.

  XX = 10   The frame is part of a fragmented frame.

  XX = 11   Reserved.

- **Block number:** The value of the block number field depends on the value of the extended frame control, XX:

  If XX = 00   Block number is not included in the frame.

  If XX = 01   Block number represents the number of fragmented frames.

  If XX = 10   Block number represents the number of transmitted frames. For example, block number 01 means the second fragmented frame, block number 02 means the third fragmented frame, and so on.

- **ACK:** The ACK field indicates the acknowledgment of a fragmented frame.
- **Destination endpoint:** The address of the destination endpoint, which may range from 0 to 240.
- **Cluster ID:** This field identifies the 8-bit cluster in a profile.
- **Profile ID:** Determines the destination profile for which the frame is intended.
- **Source endpoint:** The source endpoint number ranging from 0 to 240.
- **APS counter:** This is used to prevent duplicate frames. For any new frame transmitted by a source to a destination, this field is incremented by 1. All blocks belonging to a fragmented application data unit use the same APS counter number. For each new transaction, the APS value is set to 0.
- **Group address:** When the delivery mode is 11, this field is present in the frame.

Acknowledgment frame format: When the ACK Request field of the data frame is set to 1, the recipient of the frame should send an acknowledgment frame to the originating source. Figure 6.9 shows the acknowledgment frame format.

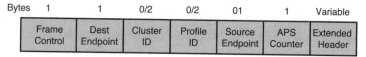

Figure 6.9   Acknowledgment frame format

6.6 APS COMMAND FRAME FORMAT

Figure 6.10 shows the APS command frame format.

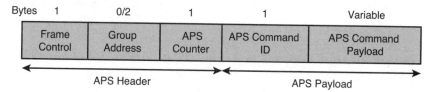

Figure 6.10   APS command frame format

- **Frame control:** See Figure 6.7 and corresponding description.
- **APS command ID:** Identifies the types of commands:

  Key-establishment command: Establishes the link key between two end devices by using Symmetric-Key Key Establishment (SKKE).

  Transport-key command: Transports a key between devices. It can use either a secured key transport through a trust center or an unsecured transport by loading the device with the initial key.

  Device update command: Updates the device's 64-bit extended and 16-bit short addresses.

  Remove device: Removes a device from the network.

  Request key: Requests a key from the coordinator.

  Switch key: Sent by a trust center to inform a device to switch its network key.

SUMMARY

This chapter presented an overview of the ZigBee APS and its various functions. The following are the key concepts described in the chapter:

- The application support sublayer (APS) is located between the application and network layers.
- It uses the APS data entity (APSDE) for the transmission and reception of data.
- It uses the APS management entity (APSME) for the transmission and reception of management primitives.
- The APS Information Base (AIB) contains constant and variable attributes.
- When the APS receives an application payload, it adds the address and control fields, and passes the resulting payload to the lower layer.
- APS frames are of the following types: data, command, and acknowledge.

REFERENCES

1. ZigBee Alliance, ZigBee Specification Document 053474r17, 2008

2. ZigBee Alliance, ZigBee Specification Document 0748855r05, 2008

3 Daintree Network, Comparing ZigBee Specification versions, www.daintree.net/resources/spec-matrix.php

4. IEEE 802.15.4 Wireless Medium Access Control (MAC) and Physical Layer (PHY) for Low -Rate Wireless Personal Area Networks, Sept. 2006

5. Ember Corp, EmberZNet Application Developer's Reference Manual, Oct. 2008

6. Document 075123r01ZB, "ZigBee Cluster Library" ZigBee Alliance Oct. 2007

CHAPTER 7

APPLICATION LAYER

INTRODUCTION

The application layer consists of application objects (endpoints), which hold user applications and ZigBee device objects (ZDOs). Figure 7.1 shows the application layer. A node can have a maximum of 240 application objects, ranging in function from home control lighting to heating and cooling of the home. The endpoints are addressed from 1 to 240; addresses 241 to 254 are reserved for future use, and address 255 is used as the broadcast address. Endpoint address 0 is assigned to the ZDO. Each endpoint can have one application profile.

7.1 APPLICATION OBJECT (ENDPOINT)

The application object holds an application profile (application program) that is developed by the user or the ZigBee Alliance. The application profile transmits and receives data through the APSDE-SAP. The ZigBee Alliance has developed several application profiles, such as home automation and smart energy. Each profile defines the type of devices that are used and describes the function of each device, such as input (switch position on and off) or output (light on and off). For example, the home automation (HA) profile contains a light sensor, a remote-control switch, and a dimmer light control.

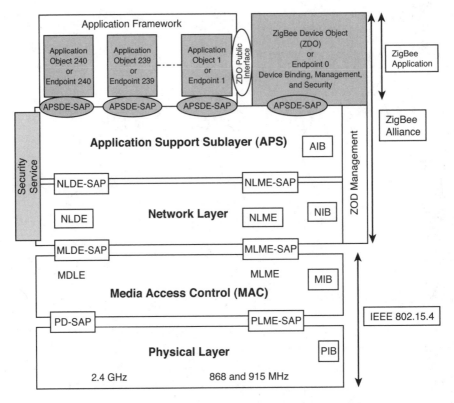

Figure 7.1 Application layer interface with the application support (APS) sublayer

7.2 ATTRIBUTE, CLUSTER, CLUSTER LIBRARY, AND PROFILE

7.2.1 ZigBee Application Profile

A profile resides in an endpoint and consists of devices and clusters that, together, form an application (for example, HA). The HA profile defines the devices and the function of each device used for home automation. Such devices include switches, dimmers, and sensors. Each profile is identified by a 16-bit number called the profile ID. Profiles are classified as being public or a manufacturer-specific private profile. Figure 7.2 shows a profile and its components.

- **Public profile:** The public profile specification is developed by a ZigBee Alliance member. Each public profile is standardized so that it can be used in any ZigBee device regardless of the manufacturer. ZigBee certifies these profiles and assigns each a profile

ID; the public profile ID range is from 0x0000 to 0x7fff. The ZigBee HA product supports E-mode commissioning. In E-mode commissioning, the device will be operational by pushing one button or through a remote device. Table 7.1 shows the ZigBee Alliance Profile names with corresponding IDs.

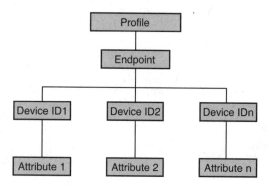

Figure 7.2   Profile and its components

Table 7.1   Public Profile Names and Corresponding Profile IDs

Profile Name	Profile ID
Industrial plant monitoring (IPM)	0x0101
Home automation (HA)	0x0104
Commercial building automation (CBA)	0x0105
Telecom application (TA)	0x0107
Personal, home, and hospital care (PHHC)	0x0108
Advance metering initiative (AMI)	0x0109

- **Manufacturer-specific profile:** A user develops his own profile and requests a profile ID from the ZigBee Alliance. The manufacturer profile ID range from 0xbf00 to 0xffff.

7.2.2 Attribute

Attributes indicate the function or data of a device connected to a node. For example, the attribute of a switch represents the position of the switch, which can be on or off. A thermostat that displays the temperature is a data attribute; each attribute is identified by a 16-bit number called the attribute identifier (ID).

- **Attribute data type:** An attribute data type is represented by 8 bits and defines the data type used to represent the attribute, such as signed 8-bit, signed 16-bit number, single-precision number, double-precision number, or character string. Table 7.2 shows the attribute code and corresponding data type.

Table 7.2 Attribute Code and Corresponding Data Type

Attribute Data Code	Data Type
0x00	No data
0x08 through 0x0B	8-, 16-, 24-, 36-bit data
0x18 through 0x1A	8-, 16-, 24-, 32-bit bitmap
0x10	Boolean
0x20 through 0x22	8-, 16-, 24-, 32-bit unsigned integer
0x28 through 0x2b	8-, 16-, 24-, 32-bit signed integer
0x30	8-bit enumeration
0x31	16-bit enumeration
0x39	Single precision
0x42	Character string

7.3 CLUSTER

A cluster is a collection of attributes and commands that is used to perform a specific function. For example, an on/off cluster is used to turn on/off a device. The command is used to manipulate attributes on the cluster. Each cluster works in the form of a client/server model. The client sends a request to the server, and the server processes the request. Each command has a sender and receiver. Therefore, a cluster consists of a server side and a client side. Figure 7.3 shows a block diagram of an on/off cluster. The cluster generated from the client is called the output cluster, and the cluster accepted by the server is called the input cluster.

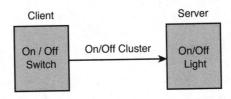

Figure 7.3 Block diagram of a cluster function

A cluster is represented by 16 bits, called the cluster ID. The ID for the on/off cluster is 0x0006. The on/off cluster can be used for any on and off device, such as garage opener, light switch, or pump. A cluster defines the format of the application support (APS) sublayer payload in the APSDE-DATA.Request or the APSDE.DATA.Indication.

7.3.1 Cluster Format

Figure 7.4 shows the general frame format of a cluster used as the payload of an APSDE-DATA.Request or APSDE.DATA.Indication.

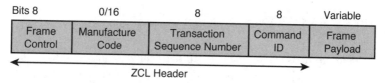

Figure 7.4   Cluster format

- **Frame Control:** 8 bits (see Figure 7.5).

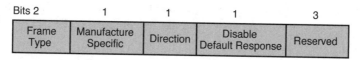

Figure 7.5   Cluster control field

- **Frame type**

  00: Indicates the command is a general cluster command (see section 7.4, "General Cluster Commands")

  01: Indicates the command is specified by the cluster, such as an on/off cluster

  10-11: Reserved

- **Manufacturer specific**

  1: Indicates that the manufacturer code is in the cluster frame

  0: Indicates that the cluster frame format does not include manufacturer code

- **Direction:** Defines the direction of the cluster frame.

  1: Indicates that the frame is from the server to the client

  0: Indicates that the frame is from the client to the server

- **Disable Default Response:** Indicates whether the receiver of the cluster is required to send a response.

  0: The receiver of the cluster is required to send a response

  1: The receiver of the cluster is not required to send a response

- **Manufacturer code:** Defines the manufacturer of the ZigBee cluster; this code is provided by ZigBee.

- **Transaction sequence number:** This field holds the sequence number of the transaction.

- **Command ID field:** The command ID field defines the type of command used by the client.

7.4 GENERAL CLUSTER COMMANDS

The general cluster commands are used when the Frame Type field described in Figure 7.5 is set to 0x00. The general cluster commands are used to manipulate, read, or report attributes based on the client/server model. Each cluster command is represented by 8 bits, called the command identifier (ID).

7.4.1 Read Attributes (ID=0x00)

The read attributes command is used to read one or more attributes from another device in the network. This is sent by the client to the server. Figure 7.6 shows read attributes command format.

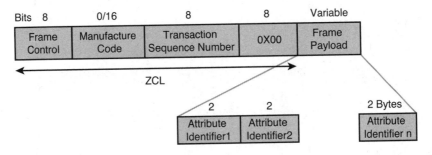

Figure 7.6 Read attributes command

7.4.2 Read Attributes Response (ID=0x01)

This command is sent by the server to the client in response to a read attributes command. Figure 7.7 shows the read attributes response format.

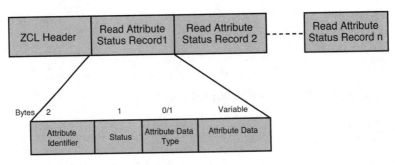

Figure 7.7 Read Attributes response format

- **Attribute Identifier (16 bits):** This is the ID of the attribute that was read.
- **Status:** This field is set to success if the attribute was read successfully.

7.4.3 Write Attributes (ID=0x03)

This command is transmitted from the client to change the value of an attribute or attributes in another device. Figure 7.8 shows the write attributes command.

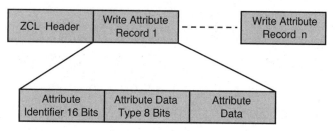

Figure 7.8   Write attributes command format

7.4.4 Write Attributes Undivided (ID=0x03)

This command is used to change all attributes of another device. If any attribute cannot be written, it does not change any of them.

7.4.5 Write Attributes Response (ID=0x04)

This is the response to a write attribute or write attributes undivided command. It is sent from the server to the client and reports the status of the write operation to the client. Figure 7.9 shows the format of write attributes response.

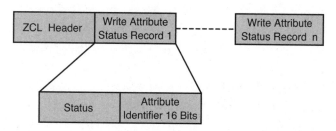

Figure 7.9   Write attributes response format

7.5 ATTRIBUTE REPORTING

Attribute reporting is used by the cluster to automatically report its attributes. This is done by configuring the cluster on the server side. The server-side cluster can report it attributes periodically or on the change of an attribute. For example, a thermostat cluster can be configured so that it periodically reports the temperature to a display panel or only reports the temperature when the temperature changes. The configure report command must be sent to the server-side cluster to enable it for automatic reporting. Automatic reporting can also be done on demand. The client periodically sends a read attributes command to the server, and the server responds with a read response.

7.5.1 Configure Report Command

The configure report command is used to configure the cluster for automatic reporting of its attributes to the client. Figure 7.10 shows the format of the configure report command.

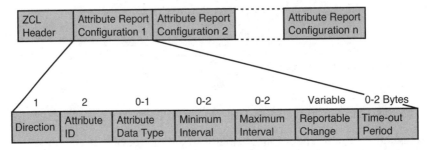

Figure 7.10   Configure report command frame format

In the ZigBee Cluster Library (ZCL), the header of the Frame Type field is set to 0b00 to indicate that the command is a general cluster command.

- **Direction:** 0x00 means the configuration command is transmitted to the server cluster and the Minimum and Maximum Interval fields are part of the frame, but the timeout period is not part of the frame. 0x01 means the record is sent to the client cluster and the Minimum and Maximum Interval fields and data type are not part of frame.
- **Attribute identifier:** If the Direction field is set to 0x00, this field contains the attribute ID (16 bits) to be configured. If the direction is set to 0x01, the attribute ID determines the attribute that must be reported to the client.
- **Minimum and Maximum Interval fields:** Each is 16 bits and represents the minimum interval and maximum interval in seconds for reporting attributes.
- **Timeout period (16 bits):** Maximum timeout between each report.
- **Reportable change:** Minimum changes that generate an immediate report.

Example

> The configuration of an on/off device (server) to automatically report every 30 seconds its on/off attribute would be as follows.
>
> - **Attribute identifier:** 0x0000.
> - **Attribute data type:** 0x01 (Boolean).
> - **Minimum interval:** 10 seconds. Do not report in less than a 10-second interval.

- **Maximum interval:** 30 seconds. Report at least every 30 seconds.
- **Reportable change:** When Boolean value changes, immediately report.

After sending the configure report command, the server responds with a configuring report response command that indicates the status of the command (for example, success, or perhaps failure due to unsupported attributes or invalid data type).

7.5.2 Report Attributes Command

When the server cluster is successfully configured, the server uses the report attributes command to report its attributes based on the attribute configuration report command. Figure 7.11 shows the report attributes command format. This command is generated when one of following conditions has been met:

- **Periodic reporting:** The periodic reporting uses the maximum time interval for its periodic reporting.
- **Change of attributes:** If any change is made to an attribute, the server transmits the report attributes command to the client. This is useful for a thermostat to transmit its temperature to a display device, for instance.

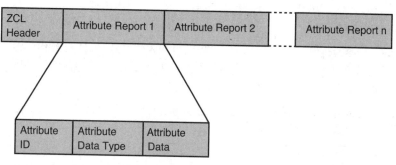

Figure 7.11   Report attributes command format

7.6 ZIGBEE CLUSTER LIBRARIES

The ZigBee Alliance developed numerous clusters and collections of clusters called the ZigBee Cluster Library (ZCL). These cluster libraries are organized based on their function domain.

- **General cluster library:** The general cluster library contains basic, on/off, level control, scenes, alarm, time, and group clusters. Table 7.3 shows some of the clusters in the general cluster library and their corresponding IDs.

Table 7. 3 General Clusters and Corresponding ID

Cluster Name	Cluster ID	Description
Basic	0x0000	Attributes related to device information such as ZCL version, application version, power source, and stack version
Power	0x0001	Attributes related to device power management
Device temperature configuration	0x0002	Attributes related to device internal temperature
Identify	0x0003	Attributes to put device in identification mode
Groups	0x0004	Attributes to configure a group of devices
Scenes	0x0005	Attributes for scene configuration
On/off	0x0006	Attributes turn a device on or off
On/off switch configuration	0x0007	Attributes to configure on/off
Level control	0x0008	Attributes for dimmer switch
Alarm	0x0009	Attributes to configure an alarm
Time	0x000A	Attributes for real time
RSSI	0x000B	Attributes for channel parameter

- **Closures:** Contains cluster for shade configuration.

- **HVAC:** Contains clusters for heating, ventilation, and air conditioning (HVAC) (for example, for pumps, thermostats, fans, and dehumidifiers).

- **Lighting:** Contains clusters for light and ballast configuration.

- **Measurement and sensing:** Contains clusters for illumination, temperature measurement, pressure, and flow measurement.

- **Security and safety:** Contains clusters for intruder alarm systems (IAS).

- **Device description ID:** Each device is represented by a 16-bit identifier. Table 7.4 shows some device names and their corresponding IDs for lighting home automation.

Table 7.4 Device Names and Correspondence Device IDs

Device Name	Device ID
On/off light	0x0100
Dimmable light	0x0101
Color dimmable light	0x0102
On/off light switch	0x0103
Dimmable switch	0x0104

7.6.1 On/Off Switch Cluster (cluster ID=0x0007)

The on and off switch attributes are divided into switch information attributes that are used to define the type of switch and switch setting attributes that define the function of the switch, as shown in Table 7.5.

Table 7.5 On/Off Switch Attribute

Switch Attribute Type	Attribute ID	Attribute Value	Attribute Data Type
Information attribute	0x0000	0x00 = Toggle switch 0x01= Momentary switch	8-bit enumeration
Setting attribute	0x0010	0x00 = On 0x01 = Off 0x02 =Toggle	8-bit enumeration

7.6.2 On and Off Cluster (cluster ID=0x0006)

The on and off cluster can be used for any device that requires on or off functionality (for example, lights or pumps). This cluster is generated by the client side. Table 7.6 shows attributes of an on/off cluster.

Table 7.6 Attributes of an On/Off Cluster

Attribute Identifier	Attribute Command ID	Type
0x0000	0x00 Off, 0x01 On	Boolean

7.6.3 Simple Application Profile

This profile defines one switch to control a light using the HA profile. In this example, the switch acts as a client and the light as a server:

1. Select a profile ID: Home automation profile ID= 0x0104.
2. Select endpoint: 0x20.
3. Select device ID: on/off switch 0x0013.
4. Select cluster for on/off switch 0x0007.
5. Information Attribute ID for switch: 0x0000.
6. Select switch type: 0x01, momentary switch.
7. Attribute data type: 0x30 (8-bit enumeration).

And for the on/off light

1. Profile ID: 0x0104.
2. Select endpoint: 0x10.
3. Select device ID: On/off light is 0x0100.

4. Select cluster for On/off light is 0x0006.

5. Attribute ID is 0x0000.

6. Attribute data type: 0x30 (8-bit enumeration).

7. Current data 0x00 for off position.

7.7 ZIGBEE DEVICE OBJECT (ZDO)

The ZDO is an application endpoint that is located in the application layer with an endpoint address of 0. As shown in Figure 7.1, the ZDO also interfaces with the APSDE-SAP (application support sublayer data entity) for data transfer and APSME-SAP (application support sublayer management entity) for control messages. The ZDO performs the following device management tasks:

- Determines the type of device in a network (for example, end device, router, or coordinator)
- Initializes the APS, the network layer, and the security service provider
- Performs device and service discovery
- Initializes the coordinator for establishing a network
- Security management
- Network management
- Binding management

7.8 ZIGBEE DEVICE PROFILE (ZDP)

The ZDP is a set of device descriptors located in the application layer. These descriptors, listed here, can be accessed by other nodes in the network:

- Device discovery
- Service discovery
- Binding and unbinding
- Network management

7.8.1 Device and Service Discovery

ZDO provides commands for the devices so that they can determine the capabilities of the other nodes in the network. Service discovery is used by an end device to discover the types of services offered by other devices in a personal-area network (PAN) and their capabilities. The device may send a broadcast service discovery or a unicast service discovery. A ZigBee node can request the capabilities of the other nodes in a network, such as device type, type of application objects, and general information about each application object. This information is stored in a node descriptor. Alternatively, the node can send a broadcast device discovery message to the other nodes in the network. They will respond to this message with their addresses. Service discovery can be used to determine the following:

- **Active endpoint:** Mandatory.
- **Node descriptor:** Mandatory.
- **Node power descriptor:** Mandatory.
- **Simple descriptor:** Mandatory.
- **Complex descriptor:** Optional.
- **User descriptor:** Optional.
- **Broadcast service discovery:** In a broadcast service discovery request, the device requesting service can receive a large amount of information from the other devices. Therefore, broadcast service discovery specifies the requested information: simple descriptor, node descriptor, power descriptor, complex descriptor, or user descriptor.
- **Unicast service discovery:** When a device requests a set of services from a specific device, the specific device always responds unless it is in sleep mode. When the device is in sleep mode, the router or coordinator associated with the device responds to the request.

Service discovery also supports the following requests.

- **Active endpoint:** The device requests the active endpoints from another device.
- **Matching simple descriptor:** The request may contain a profile ID and a list of cluster IDs, which the responder will use as criteria for selecting the list of endpoints with the same criteria for its response.

7.8.2 Node Descriptor

The node descriptor contains the following information about the node:

- **Logical type (3 bits):** 010 end device, 001 router, and 000 coordinator.
- **Complex descriptor (1 bit):** Zero means the device does not have a complex descriptor; one means the device does contain a complex descriptor
- **Frequency band of operation (5 bits):**
  00000   868–868.6MHz
  00010   902–929MHz
  00011   2.4–2.483GHz
  The rest are reserved.

- **MAC capability (8 bits):** MAC capability is represented by $B_0\ B_1\ B_2\ B_3\ B_4\ B_5\ B_6\ B_7$.

  Bit 0:   The device can be used as a coordinator or router.

  Bit 1:   The device is a full-function device (FFD).

  Bit 2:   The current power source is main power.

  Bit 3:   The receiver is on while the transceiver is idle.

  Bits 4–5: Reserved.

  Bit 6:   The device is able to send and receive secured frames.

  Bit 7:   Reserved.

- **Manufacturer code (16 bits):** Allocated by the ZigBee Alliance.
- **Maximum buffer size (8 bits):** Defines the buffer size.
- **Maximum incoming transfer size**: Defines the maximum size of the transfer unit.
- **Server mask (16 bits):** This field defines the server capability in the network:

  0x0000   The server is the primary trust center.

  0x0001   The server is the backup trust center.

  0x0002   The server holds the primary binding cache table.

  0x0003   The backup holds the primary binding cache table.

  0x0004   The primary discovery cache.

  0x0005   The backup discovery cache.

  0x0006   The server is network management.

7.8.3 Node Power Descriptor

The power descriptor defines how the node is powered:

- **Power mode:** The power mode contains the following information:

  Device receiver is on all the time.

  Device wakes up periodically as specified by the network.

  Device wakes up when application requires.

- **Available power source:** Indicates the type of power source: main power, recharge-able power, or disposable battery.
- **Current level of power source:** Indicates the level of charge in the power source.

7.8.4 Simple Descriptor

The simple descriptor defines a profile and the input/output clusters supported by the active endpoint in the node:

- The endpoint address represented by 8 bits.
- The application profile identifier represented by 16 bits.
- Application device identifier (16-bit), which defines the device description.
- The 4-bit application device version defines the version of the device description. The current version is 0000b.
- The 8-bit application input cluster count defines the number of input clusters in the endpoint.
- The application input cluster list defines the input clusters. Its size is equal to 16 times the number of input clusters.
- The 8-bit application output cluster count defines the number of output clusters.
- The application output cluster list defines the output cluster. Its size is equal to 16 times the number of output clusters.

7.8.5 Complex Descriptor

The complex descriptor offers detailed information about the device, such as the following:

- The type of character set used (e.g., ASCII, Unicode)
- Manufacturer name: name of the node manufacturer
- Model name, serial number
- Manufacturer URL

7.9 DEVICE DISCOVERY

Device discovery is used to identify other devices in a PAN. The device discovery command supports both the IEEE 64-bit and 16-bit network address and can be sent as either a broadcast or unicast message. To a broadcast discovery request, each device in the network responds with its address. In addition, any coordinator or router also responds with its list of associated devices.

To a unicast request, only the device that receives the request responds with its address. Again, if the receiving device is a coordinator or router, the response will also include the addresses of the devices that are associated with it.

7.9.1 Primary Discovery Cache Device

A network should have a primary discovery cache device for storing node descriptors of the devices that are in sleep mode. The primary discovery cache device can be a router or a coordinator. It is the primary discovery cache device that responds when the requested device is in sleep mode. Before any device goes into sleep mode, it transmits the following information to the primary discovery cache:

- IEEE address
- Network address
- Active endpoints (list of)
- Simple descriptor
- Power descriptor

7.9.2 Service and Device Discovery Commands

The device profile operates based on the client/server model: The client sends a request to the server, and the server responds to the request. The following are some client requests that are used for device and service discovery.

- **Network address request:** This is generated by a device in the network that is requesting the 16-bit network address of a remote device by transmitting the IEEE address of the remote device. This command is broadcast. The remote device compares the supplied IEEE address with its IEEE address. If they match, the remote device responds with the network address response command, which contains the 16-bit network address.

- **IEEE address request:** This command, sent by a local device to a remote device, requests the IEEE address of remote device using the 16-bit network address of the remote device. The remote device responds with its IEEE address and 16-bit network address.

- **Node description request:** The device request sends a unicast packet requesting the node description of the remote device.

- **Power description request:** The local device requests the power description of the remote device using unicast addressing.

- **Active endpoint request:** The local device requests information about the active endpoints in the remote device. An active endpoint is an endpoint with an application profile described by the simple descriptor. This request can be transmitted using unicast or broadcast addressing.

- **Simple descriptor request:** A local device requests a simple descriptor in a specific endpoint of a remote device.

- **Match descriptor request:** The local device transmits a request that includes the following criteria for matching: network address, profile ID, number of input clusters, input cluster IDs, number of output clusters, and output cluster IDs. This request can be sent to a remote device using unicast or broadcast addressing.

- **Complex descriptor request:** The local device requests a complex descriptor from the remote device by using unicast addressing.

- **User descriptor request:** The local device requests a user descriptor from the remote device by using unicast addressing.

- **Device cache request:** This request, sent as a broadcast message from a device, locates the primary cache device on the network. Only the device that is designated as the primary cache responds to the request by sending a success message. The local device then sends a discovery storage request to the primary discovery cache to determine whether the primary cache has sufficient memory.

- **End device announcement:** A device that has joined as an end device or router, or a device rejoining the network, sends a broadcast announcement to the devices in the network. The announcement contains the network address of the device, IEEE address, and a MAC capability flag bit.

- **User destination set request:** This command is used by the local device to configure the user description of a remote device.

- **Server discovery request:** The local device sends a broadcast server discovery to locate a server in the network.

- **Discovery cache storage:** The local device sends this request to the primary cache device to determine whether the primary cache has enough memory. The request command contains the following information:

  Device's network address

  Device's IEEE address

  Size of the node descriptor (bytes)

  Power descriptor size

  Active endpoint size

  Simple descriptor list size

  If the primary cache device has enough memory space, it will respond with success and store the above information in its cache. If it does not have enough memory, it sends a message that there is not enough memory space.

- **Store node descriptor:** The local device requests that the primary cache store its node descriptor.

- **Store power descriptor:** The local device requests that the primary cache store its power descriptor.

- **Store active endpoints request:** This is sent by the local device to the primary cache device to store the active endpoints of the device. The following information will be sent to the primary cache for storage:

Network address

IEEE address

Number of active endpoints

List of active endpoints (8-bit address)

- **Store simple descriptor request:** This is sent by the local device to the primary cache device for storing the simple descriptor of the device. The local device will send the following information to the primary cache:

Network address

IEEE address

Length of the simple descriptor in bytes

Simple descriptor

- **Remove node cache request:** The device requests that the primary cache remove the device from the discovery cache.

7.10 BINDING

Binding is a process used to establish a logical relation between two endpoints located in different devices. A switch that controls a light is bound to the light, for instance. Binding is unidirectional. The switch generates an output cluster that becomes the input cluster for the light. The binding between two devices requires the following information:

- The source node address and the source endpoint address
- The destination node address and the destination endpoint address
- The cluster ID and the profile ID of the applications

The following types of binding are supported by ZigBee:

- **One-to-one binding:** One switch controlling a single light
- **One-to-many binding:** One switch controlling several lights
- **Many-to-one binding:** Multiple switches controlling a single light

7.10.1 Binding Process

Figure 7.12 shows a home lighting control network. In this figure, the bedroom lights are controlled by the bedroom switch with EP1 (endpoint 1). The bedroom switch, EP1, is bound to EP2, EP3, and EP4. The binding information is stored in a binding table in the source node (switch), as shown in Table 7.7. When the switch L1 activates, the switch position sends three APS-DATA.Requests: one to the bedroom light1, one to the bedroom light2, and one to bedroom light3.

Table 7.7 Binding Table for Figure 7.17

Source Address	Source Endpoint	Cluster ID	Destination	Destination Endpoint
L1 switch	EP1	0x0006	Bedroom Light1	EP2
L1 switch	EP1	0x0006	Bedroom Light2	EP3
L1 switch	EP1	0x0006	Bedroom Light3	EP4

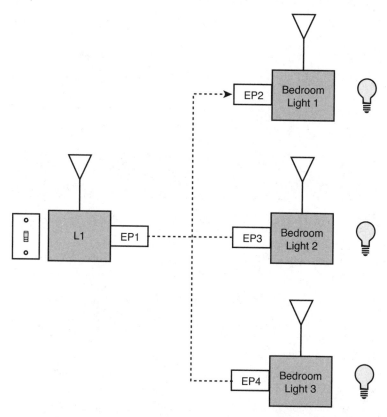

Figure 7.12 Home lighting control

In the binding process, when two devices are bound together, they share a cluster called the link cluster. In general, the binding table maps the source endpoint to the destination endpoint. For example, if a light switch controls several lights, as shown in Figure 7.12, the mapping function will be as follows.

$$As, Es, Cs = (Ad_1, Ed_1), (Ad_2, Ed_2), \ldots (Ad_i, Ed_i)$$

- **As:** The source address of the node (16-bit short address or 64-bit IEEE address)
- **Es:** The source address of the endpoint (1 through 240)
- **Cs**: Link cluster ID (8-bit)

- **Ad$_i$:** Destination address of the node
- **Ed$_i$:** Destination address of the endpoint

The application layer of a device sends an APSME-BIND.Request to its local binding table to request the addition of a record. Figure 7.13 shows the format of the APSME-BIND.Request. The local APS will notify the application layer of the outcome of the bind request by sending an APSM-BIND.Confirm. Figure 7.14 shows the APSME-BIND.Confirm format.

Source Address	Source Endpoint	Cluster ID	Dest Address Mode	Dest Address	Dest Endpoint

Figure 7.13   APSME-BIND.Request

Status	Source Address	Source Endpoint	Cluster ID	Dest Address Mode	Dest Address	Dest Endpoint

Figure 7.14   APSME-BIND.Confirm

7.10.2 End Device Binding, Unbinding, and Client Binding Management

One function of the ZDO is binding management. The ZDO provides binding and unbinding of devices, and storing of the binding and backup binding tables. All management binding commands are unicast and are as follows:

- **End device bind request command:** This command is used to bind two nodes together. The commands are sent by the source devices to the coordinator to request binding with each other. When the coordinator receives this command, it performs the following tasks. See Figure 7.15 for the end device bind request command format.

  1. When the coordinator receives the bind request, it checks whether the endpoint number is in the range of 1 to 240.

  2. If the endpoint is not between 1 and 240, the coordinator sends an end device bind response command to the source device with a status of invalid endpoint.

  3. If the end point is within range, the coordinator holds the first request for a predefined length of time and waits for the second device to issue an end device bind request command.

  4. If, during this time, the coordinator does not receive the end device bind request from the second device, the coordinator sends an end device bind response command with a status of timeout to the source device.

5. If, during this time, the coordinator does receive the end device bind request from the second device, it checks the validity of its endpoint number.

6. It then checks that the profile ID, input cluster ID, and output cluster ID of the first device match with the second device.

7. If the result of step 6 shows there is no match, the coordinator sends an end device bind response to both devices with a status of no match.

8. If the result of step 6 shows there is a match, the coordinator sends an end device bind response to the both devices with a status of success.

Binding Target	Source IEEE Address	Source Endpoint	Profile ID	Number of Input Clusters	List of Input Clusters	Number of Output clusters	List of Output Clusters

Figure 7.15   End binding request format

- **Binding Target:** 16-bit network address of the target device for binding
- **Source IEEE Address:** 64-bit network address of the requesting device
- **Source endpoint:** 8 bits
- **Profile ID:** 16-bit profile identifier for the endpoint
- **Number of input cluster**
- **Input cluster:** The size of this field equal to the size of each cluster representation (2 bytes) times the number of input clusters.

7.10.3 Bind and Unbind Commands

The following are the bind and unbind client requests. Both the requests and the responses are optional:

- End device bind request
- Bind request
- Unbind request
- Bind register request
- Replace device request

Unbind request: A device requests the removal of an entry from the binding table. The unbind request command contains the following information:

- Source address
- Source endpoint
- Cluster ID
- Destination address mode
- Destination endpoint

Local binding register request: This command is sent by a device to notify the primary binding table cache that the device will store its own binding table.

Replace device request: This is issued by a device to the primary binding table to replace all entries in the binding table.

Back-up binding entry table: This is generated by the primary binding table cache and sent to a remote backup binding table for storing the backup table.

Recover binding table: This is issued by the primary binding table cache and sent to the remote backup for restoring the binding table.

7.11 NETWORK MANAGEMENT COMMANDS

The device profile supports the following network management commands. The network management command is based on a client/server model. The client requests and the remote device respond to the client request. The following commands are used for network management:

- **Network discovery request:** A local device issues this command to a remote device for the execution of an NLME-NETWORK.Discovery. The remote device reports back the result of the networks found in its vicinity.
- **Link quality indication (LQI) request:** The local device sends this request to the coordinator or router requesting a neighbor list with the LQI of each neighbor.
- **Routing discovery management request:** A local device requests the routing table from a remote device.
- **Leave request:** The local device (coordinator or router) sends this command to a remote device to remove it from the network; when the remote device receives this request, it executes an NLME-LEAVE.Request command.
- **Permit joining request:** A local device (coordinator or router) sends this command to a remote device to permit it to join the network directly. The remote device then executes an NLME-DIRECT-JOIN.Request.
- **Management cache request:** The local device sends this command to request a list of end devices that are registered with the primary cache device.
- **Power description store request:** Request to store the power description of the device on the primary discovery cache device.
- **Active endpoint store request:** The end device requests the storing of its list of active endpoints in the primary discovery cache.
- **Remove node cache request:** Sent from a device to the primary discovery cache for the removal of the device from the cache.
- **Management network discovery request:** This command, sent from the source device to the remote device, contains a list of channels that the remote device will scan

by executing an NLME network discovery request command. The remote device sends the result of the scan to the source device using a management discovery response.

- **Management link quality request:** This is used by the source device to request the LQI value of its neighbor. The neighbor nodes should be a router or ZigBee coordinator.

  The ZigBee coordinator or router reads information from its network neighbor table, which is stored in the NIB, by using NLME-GET request and sends the result to the source device by using the management LQI response command.

- **Management routing request:** This command is sent by the end device to the router or ZigBee coordinator for the purpose of requesting the routing table entries.

- **Management leave request:** This is sent from the management device to request that a remote device leave the network.

- **Management join request:** The management device requests a remote device to join the network.

7.12 ZIGBEE COORDINATOR STARTUP

During startup, a device accesses those ZigBee startup parameters listed here, which are necessary for a device to become a coordinator:

- **Network extended PAN ID:** This is the PAN ID of the network that the device is joining. If the PAN ID is set to 0, the device is not connected to a network.

- **APS designated coordinator (true/false):** True means the device will become a coordinator during startup.

- **APS channel mask:** List of channels that the device can choose from for operation.

- **APS use extended PAN ID:** If the APS designator is false, the APS extended PAN ID is used to join the network.

- **APS use insecure join (true/false):** Indicates whether to join the network using an insecure process.

SUMMARY

This chapter presented the functions of the ZigBee application layer:

- The application layer consists of the application endpoints and the ZigBee device object (ZDO).

- Each node can have up to 240 application endpoints, which are represented using an 8-bit address.

- The function of the application endpoint is to transmit data and receive data.
- Information generated by the application endpoint is called an attribute. For example, the switch positions, on and off, are attributes for the application endpoint.
- A cluster is collection of attributes and commands.
- The attribute identifier defines the attribute in a cluster and is represented by 16 bits.
- A cluster works in the form of a client/server model
- Each cluster is represented by a 16-bit ID, called a cluster ID.
- Cluster commands read and write attributes, read and write responses, and report attributes.
- The collection of clusters is called the ZigBee Cluster Library (ZCL).
- A general cluster library contains basic, on/off, level control, scenes, alarm, time, and group clusters.
- A profile consists of an endpoint, device ID, and cluster, which, together, form an application.
- Profile IDs, cluster IDs, and attribute IDs are all represented by 16 bits.
- The functions of the ZDO are service and device discovery, initialization of the coordinator, security management, binding management, and network management.
- The service discovery is used by a device to discover types of services offered by other devices in same network.
- Each end device has a node, power, simple, and complex descriptor.
- The node descriptor contains the type of device, frequency band operation, manufacturer code, maximum buffer, and MAC layer capability.
- The power descriptor contains the power mode and the available power source.
- The simple descriptor contains information about each endpoint: endpoint address, application profile, device descriptor ID, and the number of input/output clusters.
- The complex descriptor contains detailed information about a device: the manufacturer name, model name, and serial number of the device.
- Binding is a process that establishes a logical relation between two endpoints.
- Binding can be one to one, one to many, or many to one.
- The ZDO uses device and service discovery commands to discover details about the other devices.
- ZDO binding management uses binding commands to bind and unbind endpoints.
- The ZDO uses network management commands for network discovery, routing discovery, link quality indication (LQI), leave requests, and join requests.

REFERENCES

1. Document 053474r17, ZigBee Specification, ZigBee Alliance, October 2007

2. Document 075123r02, ZigBee Cluster Library Specification, ZigBee Alliance, May 2008

3. Document 053520r25, Home Automation Profile Specification, ZigBee Alliance, October 2007

4. Document 075356r14ZB_AMI_PTG-AMI_Profile_Specification.pdf, ZigBee Alliance, May 2008

5. Drew Gibson, ZigBee Wireless Networking, Newnes, 2008

6. www.radiopulse.co.kr, Profile and ZLC

CHAPTER 8

ZigBee Security

INTRODUCTION

Network security plays an important role in successful transmission of information between devices. Currently, many people are accessing their bank accounts, buying and selling stocks, and paying bills over the Internet. People using these services need to have their transactions secured. This means that the information transmitted should not be accessible or modified by anyone other than the authorized user. Network security is implemented to protect information in transit and to protect a system from an attack.

8.1 ELEMENTS OF NETWORK SECURITY

The following elements of network security are necessary to ensure a secure network:

- **Secrecy or confidentiality:** Secrecy provides privacy and protects information from being intercepted. Even if someone intercepts the information, the information would have no meaning.
- **Authentication:** Authentication methods verify the identity of a person or computer accessing the network.
- **Integrity:** Integrity maintains data consistency and prevents tampering with information.
- **Nonrepudiation:** Nonrepudiation provides proof of origin to the recipient.

8.2 INTRODUCTION TO CRYPTOGRAPHY

Cryptography is the analysis and deciphering of codes and ciphers. In computer networking, cryptography is the science of keeping information secure. The process of encoding and decoding information is called encryption and decryption, respectively. Figure 8.1 shows a cryptographic model.

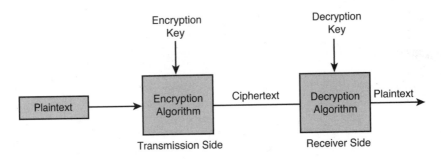

Figure 8.1 Cryptographic model

In this model, the message is referred to as plaintext or cleartext, and the encrypted text is called the ciphertext. The plaintext is encrypted by an encryption algorithm and an encryption key. The ciphertext is then transmitted over the communication channel. At the receiving side, the ciphertext is decrypted by a decryption algorithm and a decryption key resulting in plaintext. The encryption and decryption algorithms are called ciphers. Cryptanalysis is the art of breaking ciphers or decrypting information without having the key.

Cryptography can be divided into two classes: classical and modern cryptography. Classical cryptography was used in the early days for noncomputing applications; for example, character substitution is a form of classical cryptography. Modern cryptography is now used for modern data communication. The types of modern cryptography are symmetric key cryptography (private key or secret key) and public key cryptography.

8.2.1 Symmetric Key Cryptography

In symmetric cryptography, the transmitter and receiver of a message share a key for encryption and decryption of data, as shown in Figure 8.2.

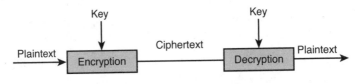

Figure 8.2 Block diagram of symmetric cryptography.

Plaintext is encrypted using an encryption algorithm and a key, resulting in cipher text, which is then transmitted to the receiver. The receiver uses the same key as the transmitter to decrypt the ciphertext into plaintext. The advantage of symmetric cryptography is that it is simple and fast. The disadvantage is that the transmitter and the receiver have to exchange keys.

The symmetric algorithm can be divided into stream ciphers and block ciphers.

Stream Ciphers

The plaintext is encrypted one bit at a time, similar to the one-time padding method. In the one-time padding method, "padding" text is converted to binary, and the binary string is then XORed with another binary string (the key), which results in the ciphertext.

Block Ciphers

A block cipher is a group of bits that are encrypted simultaneously. For example, the Data Encryption Standard (DES) operates on blocks of 64 bits. Some of the classical block cipher encryptions are substitution and transposition.

8.2.2 Advanced Encryption Standard (AES)

AES or Rijndael (the combination of the names Rijman and Daemen, the developers of the AES algorithm) is a block cipher that can use a 128-, 192-, or 256-bit key for encryption. ZigBee security uses a 128-bit encryption key. The ciphertext and key are each represented by a 4 * 4 array of bytes (16 bytes * 8 = 128 bits) called the cipher state and key, respectively. Figure 8.3 shows the cipher state and key

$K_{0,0}$	$K_{0,1}$	$K_{0,2}$	$K_{0,3}$
$K_{1,0}$	$S_{1,1}$	$K_{1,2}$	$K_{1,3}$
$K_{2,0}$	$K_{2,1}$	$K_{2,2}$	$K_{2,3}$
$K_{3,0}$	$K_{3,1}$	$K_{3,2}$	$K_{3,3}$

$S_{0,0}$	$S_{0,1}$	$S_{0,2}$	$S_{0,3}$
$S_{1,0}$	$S_{1,1}$	$S_{1,2}$	$S_{1,3}$
$S_{2,0}$	$S_{2,1}$	$S_{2,2}$	$S_{2,3}$
$S_{3,0}$	$S_{3,1}$	$S_{3,2}$	$S_{3,3}$

Figure 8.3   Representation of ciphertext and key

The following steps describe the AES encryption process:

1. **Substitution step:** Each byte in the state is replaced by another byte using an S-box (Table 8.1). For example, if the current byte state is 0x19, it is replaced with the number located at row 0x10 and column 0x09, which is 0xd4.

Table 8.1 S-Box Table Used in AES Substitution Step

	0	1	2	3	4	5	6	7	8	9	a	b	c	d	e	f
00	63	7c	77	7b	f2	6b	6f	c5	30	01	67	2b	fe	d7	ab	76
10	ca	82	c9	7d	fa	59	47	f0	ad	d4	a2	af	9c	a4	72	c0
20	b7	fd	93	26	36	3f	f7	cc	34	a5	e5	f1	71	d8	31	15
30	04	c7	23	c3	18	96	05	9a	07	12	80	e2	eb	27	b2	75
40	09	83	2c	1a	1b	6e	5a	a0	52	3b	d6	b3	29	e3	2f	84
50	53	d1	00	ed	20	fc	b1	5b	6a	cb	be	39	4a	4c	58	cf
60	d0	ef	aa	fb	43	4d	33	85	45	F9	02	7f	50	3c	9f	a8
70	51	a3	40	8f	92	9d	38	f5	bc	b6	da	21	10	ff	f3	d2
80	cd	0c	13	ec	5f	97	44	17	c4	a7	7e	3d	64	5d	19	73
90	60	81	4f	dc	22	2a	90	88	46	ee	b8	14	de	5e	0b	db
a0	e0	32	3a	0a	49	06	24	5c	c2	d3	ac	62	91	95	e4	79
b0	e7	c8	37	6d	8d	d5	4e	a9	6c	56	f4	ea	65	7a	ae	08
c0	ba	78	25	2e	1c	a6	b4	c6	e8	dd	74	1f	4b	bd	8b	8a
d0	70	3e	b5	66	48	03	f6	0e	61	35	57	b9	86	c1	1d	9e
e0	e1	f8	98	11	69	d9	8e	94	9b	1e	87	e9	ce	55	28	df
f0	8c	a1	89	0d	bf	e6	42	68	41	99	2d	0f	b0	54	bb	16

2. **Shift row step:** Performs the following operation on each row of the result of step 1, as shown in Figure 8.4.

No shift on first row.

Circular shift left 1 byte (each bit is shifted 8 times) on the second row.

Circular shift 2 bytes (each bit is shifted 16 times) on the third row.

Circular shift 3 bytes (each bit is shifted 24 times) on the fourth row.

$S_{0,0}$	$S_{0,1}$	$S_{0,2}$	$S_{0,3}$
$S_{1,1}$	$S_{1,2}$	$S_{1,3}$	$S_{1,0}$
$S_{2,2}$	$S_{2,3}$	$S_{2,0}$	$S_{2,1}$
$S_{3,3}$	$S_{3,0}$	$S_{3,1}$	$S_{3,2}$

Figure 8.4 Result of shift process

3. **Mix column:** As shown in Figure 8.5, the result of step 2 is represented by matrix S, where the given matrix is multiplied by each column of matrix S to produce matrix A.

$$\begin{vmatrix} 2 & 3 & 1 & 1 \\ 1 & 2 & 3 & 1 \\ 1 & 1 & 2 & 3 \\ 3 & 1 & 1 & 2 \end{vmatrix} \times \begin{vmatrix} S_{0,0} & S_{0,1} & S_{0,2} & S_{0,3} \\ S_{1,1} & S_{1,2} & S_{1,3} & S_{1,0} \\ S_{2,2} & S_{2,3} & S_{2,0} & S_{2,1} \\ S_{3,3} & S_{3,0} & S_{3,1} & S_{3,2} \end{vmatrix} = \begin{vmatrix} A_{0,0} & A_{0,1} & A_{0,2} & A_{0,3} \\ A_{1,0} & A_{1,1} & A_{1,2} & A_{1,3} \\ A_{2,0} & A_{2,1} & A_{2,2} & A_{2,3} \\ A_{3,0} & A_{3,1} & A_{3,2} & A_{3,3} \end{vmatrix}$$

Figure 8.5   Mix column operation in AES

Where

$$A_{0,0} = 2S_{0,0} + 3S_{1,1} + S_{2,2} + S_{3,3}$$
$$A_{0,1} = 1S_{0,1} + 2S_{1,2} + 3S_{2,3} + S_{3,0}$$
$$A_{0,2} = S_{0,2} + S_{1,3} + 2S_{2,0} + 3S_{3,1}$$
$$A_{0,3} = 3S_{0,3} + S_{1,0} + S_{2,1} + 2S_{3,0}$$

4. **Add round key:** Each element of the 4 * 4 array, A, resulting from step 3 is XORed with its corresponding element in the 4 * 4 array representing the key resulting in array B, as shown in Figure 8.6.

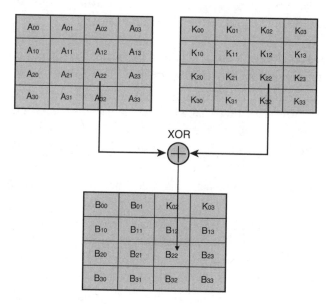

Figure 8.6   XOR operation of array A with array K

5. The result of step 4, array B, becomes the new state for step 1, and this process is repeated ten times.

8.3 ZIGBEE SECURITY

ZigBee security provides message integrity, authentication, freshness, and privacy for a ZigBee device. It offers different levels of security and uses a standard security specification called counter mode encryption plus cipher block chaining message authentication code (CCM) with the 128-bit AES (Advanced Encryption Standard) algorithm.

ZigBee security uses counter mode (CTR) with 128-bit AES for the encryption of messages and cipher block chaining (CBC) with 128-bit AES for the generation of the message integrity code (MIC).

ZigBee security uses symmetric key for all levels of security. ZigBee security can apply cryptography and frame integrity to the application and network layers. The application developer must decide at which level to apply security. Any layer that generates a frame is responsible for its security. For example, the application layer uses a link key for security, and the broadcast frame is secured using the network key.

8.3.1 ZigBee Security Keys and Trust Center

ZigBee defines three types of keys for security: link, network, and master. It also defines a trust center to manage network security and key distribution:

- **Link key or secret key:** A link key is used by the application protocol sub-layer (APS) to secure application data between two devices. There are two types of link keys:
  - *Application link key*: This key is used for security of application data between two devices and it is shared only between two devices. An application link key may be preconfigured, distributed by a trust center to the devices, generated from a master key, or installed using the Symmetric-Key Key Establishment (SKKE) protocol.
  - *Trust Center link key:* This key is used by the trust center and devices on the network for securing communication between the trust center and devices. This key is preconfigured in the devices.
- **Network key:** The network key is usually used for broadcast messages and frames generated by the network layer. Because all the devices on the PAN must be able to decipher these broadcast messages, they all share a common network key. This key is installed by the manufacturer or transported by the trust center.
- **Master key:** The master key may be installed by the manufacturer of the device or installed by a trust center. The master key is an optional key and is used for the generation of the link key.
- **Trust center:** The trust center is a dedicated network device or coordinator that is trusted by the other devices on the network. It is used to store a list of devices, master keys, link keys, and network keys, and it performs key update and device authentication for joining the network. The trust center periodically updates the network key used

by all devices on the network. It does this by encrypting the new network key with the current network key and broadcasting the key to all devices on the network. Finally, it sends a request to all devices to switch to the new key.

8.4 ZIGBEE SECURITY MODES

ZigBee defines the following security modes:

- **Standard mode:** Standard security mode is used by ZigBee and ZigBee PRO.
- **High-security mode:** High-security mode is used by ZigBee PRO.

8.4.1 Standard Security

ZigBee and ZigBee PRO support standard security, which uses two network keys that are switched from active to secondary by the trust center for encryption and decryption. Standard mode also offers optional security at the application level; node-to-node communications can use a link key for encryption and decryption of application data.

ZigBee PRO provides security enhancement for a device in sleep mode. When a device is in sleep mode, it is unable to receive the new network key from the trust center. If this occurs, when the device wakes up, it cannot communicate with other devices in the network using the old network key. The device uses the trust center link key to request a new network key from the trust center. In this mode, the "network all fresh" attribute in the Network Information Base (NIB) is set to false, which means the network layer is not required to check the incoming frame for freshness.

Function of the Trust Center in Standard Mode

The trust center performs the following functions:

- The trust center and devices in the network use the trust center link key to communicate with each other, as follows:
  - If a device requests an update of the network key, the trust center encrypts a new network key with the trust center link key and transmits it to the device.
  - When a new device joins the network, the trust center transmits an active network key to the device using the primitive APSME-TRANSPORT-KEY.Request, which is encrypted using the trust center link key.
- The trust center can generate the application link key between two devices. All devices in the network are preprogrammed with the trust center link key, and they use this link for communication with the trust center, as shown in Figure 8.7. Device A requests the

link key for device B from the trust center (the request is encrypted by the trust center link key), and device B requests the link key for device A from the trust center. (The device B request is also encrypted by the trust center link key.) The trust center generates the link key and encrypts it with its key and then transmits the result to the devices: A and B.

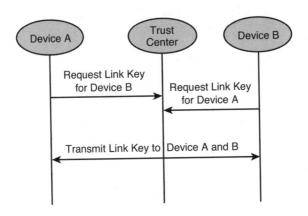

Figure 8.7 Trust center generating link key

Joining a Network in Standard Mode

In standard mode, when a device transmits the MLME-ASSOCIATE.Request primitive, the security capability field is set to 0, which indicates standard security. If this bit is set to 1, the device has the capability for high-security mode.

The following process describes how a device joins a network in standard security mode:

1. If the device does not have a network key, the following process takes place:
   a. The device sends a MAC association request command to join the parent.
   b. If the parent accepts the device, it sends success to the child and the child joins the network as an unauthenticated device.
   c. The parent of the device uses its network key and encrypts the ASME-UPDATE-DEVICE.Request and transmits it to the trust center.
   d. If the trust center allows the new device to join the network, there are two scenarios:
   i. If the device does not have the link key, the trust center transmits an encrypted network key to the parent of the device. The parent of the device then sends an unencrypted network key to the device.
   ii. If the device has the trust center link key, the trust center encrypts the network key using the trust center link key and transmits it to the device.

2. If the trust center did not accept the device, it sends an APSME-REMOVE-DEVICE request to the device's parent requesting the removal of the device from network.

Rejoining a Network:

Suppose a device was once part of a network but, because of the following cases, cannot now communicate with the network:

- (Perhaps because) The device was asleep during the update of the network key.
- (Or) The device lost its parent due to movement or an obstacle in the network.

In either scenario, the device can rejoin the network by either a secured or an unsecured rejoin process:

- **Secured rejoin process:** The device encrypts its rejoin request command (NLME-JOIN.Request with the Rejoin Network field set to 0x02) using its old network key and transmits the result to its parent. If the parent has the same network key, it informs the trust center of the rejoin by transmitting an ASME-UPDATE-DEVICE.Request to update the list of devices in the network. The parent then responds to the device by sending a network rejoin response encrypted using the network key.
- **Unsecured rejoin process:** When a device does not have the network key, it can join the network using the unsecured rejoin process, as follows:
  1. The device sends an NLME-JOIN.Request (with the Rejoin Network field set to 0x02) to its neighbor.
  2. The parent of the device sends an ASME-UPDATE-DEVICE.Request to the trust center and a network rejoin response to the device.
  3. If the trust center accepts the device, it encrypts the network key using the trust center link key and transmits it to the device.

8.4.2 High–Security Mode

High-security mode uses two network keys and separate link keys for communication between devices. The network keys, master keys, and list of devices are all stored in the trust center. The link keys are generated using the SKKE protocol. High-security mode also provides authentication. In this mode, the "network all fresh" attribute is set to true. The high-security mode provides all functions supported by standard security and the following additional functions:

- **Entity authentication:** This is used by two devices to authenticate each other based on their active network key

- **Permission table:** This table indicates which devices have permission for using commands, such as

  Permission to start up a network, permission to join or leave

  Binding and grouping

  Required to use link key

- Generating link key between devices using SKKE protocol

8.5 SECURITY MANAGEMENT PRIMITIVES

The following primitives are used for security management:

- **APSME-ESTABLISH-KEY:** This primitive is used to establish a link key between two devices using the SKKE protocol.
- **APSME-TRANSPORT-KEY:** Used for transporting the key from one device to another device. Supports request and indication primitives. Figure 8.8 shows the APSME-TRANSPORT-KEY.Request primitive format.

Destination Address	Key Type	Key Data

Figure 8.8 TRANSPORT-KEY.Request primitive format

- **Destination address:** The 64-bit IEEE address of the destination.
- **Key type:** The type of key being transported by the primitive.

Key type value = 0x00	Trust center master key, which is used to set up link keys between a device and the trust center
Key type value = 0x01	Standard network key, which is used for all devices on the network
Key type value = 0x02	Application master key, which is used to generate the link key for two devices
Key type value = 0x03	Application link key
Key type value = 0x04	Trust center link key, which is used for securing communication between devices and the trust center
Key type value = 0x05	High-security network key, which is used for high-security mode

- **Key data:** Holds the information about the key, which varies depending on the key type value.

  Key types 0x00 and 0x04 are trust center master and link keys. The Key Data field contains the parent address and trust center master or link key.

Key types 0x01 and 0x05 are for the standard and high-security network key. The information in the Data field is for the network key, and includes the key sequence number, network key, and use parent Boolean (true means key transported to the parent device, false means key transported to the device).

Key types 0x02 and 0x03 are used for the application master key and application link key. The information contained in key data is the parent address, the initiator Boolean, which indicates a destination request, and the key value.

- **APSME-UPDATE-DEVICE:** The trust center uses this primitive to update the key of a device.

- **APSME-REMOVE-DEVICE:** This is used by the trust center and parent to remove a device from the network because the device failed to authenticate.

- **APSME-REQUEST-KEY:** This is generated by a device to request an active network, master, or link key from the trust center.

- **APMS-SWITCH-KEY:** This is generated by the trust center and issued to the devices to switch to a new active network key.

- **APSME-AUTHENTICATE:** This primitive is used by a device to authenticate another device based on a shared key.

8.6 COUNTER (CTR) MODE ENCRYPTION

Figure 8.9 shows the block diagram of a CTR mode operation, where T1 is a count value generated by a counter, $T2 = T1 + 1$, $T3 = T2 + 1$ and $Tn = T(n - 1) + 1$. The message is divided into blocks D1, D2...Dn. The T1, T2...Tn are encrypted using a 128-bit AES encryption key (EK), and the results I1, I2...In are XORed with D1, D2...Dn, respectively. The output P1, P2...Pn is the encrypted message.

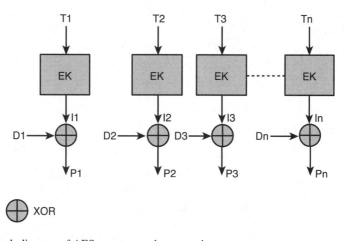

Figure 8.9   Block diagram of AES counter mode encryption

8.7 CIPHER BLOCK CHAINING (CBC) MODE ENCRYPTION

Figure 8.10 shows the block diagram of CBC mode encryption. The message is divided into 128-bit blocks. The first block, D1, is encrypted using the 128-bit AES EK, and the ciphertext, I1, is XORed with the next block of data, D2. This process is repeated until the message is fully encrypted.

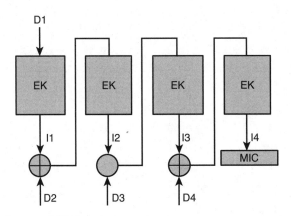

Figure 8.10   Block diagram of CBC mode encryption

In Figure 8.8, D1, D2…Dn are data blocks, and I1, I2…In are encrypted messages. In is the message integrity code (MIC); so, in Figure 8.10, I4 is the MIC.

8.8 NETWORK LAYER SECURITY

Network layer security applies security to the frame generated at the network or higher layer. The network layer can also accept secured incoming frames. If security is enabled in the network Frame Control field (see Figure 5.12 of Chapter 5, "Network Layer") or nwk-SecurityAllframes is set to true in the NIB, security must be applied to the network frame. The security level is determined by the nwkSecurityLevel attribute, which contains a security-level identifier. Table 8.2 lists all the possible security-level identifiers and their properties.

Table 8.2   Security-Level Identifiers

Security-Level Identifier	Security	Data Encryption	MIC Length in Bytes
00	None	Off	0
01	MIC-32	Off	4
02	MIC-64	Off	8
03	MIC-128	Off	16

Table 8.2   Continued

Security-Level Identifier	Security	Data Encryption	MIC Length in Bytes
04	Encryption	On	0
05	Encryption and MIC-32	On	4
06	Encryption and MIC-64	On	8
07	Encryption and MIC-128	On	16

8.8.1 Network Layer Security Attributes

The following attributes are stored in the NIB for use by the NLME in securing the network frames:

- **Network security level:** The security level of the outgoing frame (see Table 8.2). The default value is 0x05.

- **Network security material set:** This contains the following information for each material set:

  The key sequence number of the network key that was assigned by the trust center; this is used for updating the key.

  Outgoing frame counter, used to determine freshness.

  Incoming frame counter.

  The actual value of the key (16 bytes).

  Key type: 01 means standard and 05 high security.

- **Network active key sequence number:** The sequence number of the active key, used to determine the network security material set.

- **Network all fresh (true/false):** If it is true, incoming frames must be checked for freshness; otherwise, no check for freshness is made.

- **Network security all frames (true/false):** If it is true, network security must be applied to all outgoing and incoming network frames.

8.8.2 Securing Network Layer Broadcast Frames

One of the functions of the network layer is route discovery. The route discovery (NLME-ROUTE.Request) command is broadcast to all neighboring devices, causing each device to respond to the route request. The network layer uses the link key to secure (encrypt and decrypt) these broadcast network layer messages. If the network layer does not have a link key, it will instead use the network key to secure the broadcast commands.

If the "network security all frames" attribute is set to true, all outgoing frames must be secured by the network layer. The network layer uses the network active sequence number stored in the NIB to access the network security material set.

8.8.3 Applying Security to a Network Layer Outgoing Frame

If the security bit in the network Frame Control field is set to 1, the network layer management entity (NLME) performs the following actions to secure the outgoing frame:

1. Finds nwkActiveKeySeqNumber within the NIB.
2. Uses the active key sequence number to find the network security material set.
3. Encrypts the network layer payload.
4. Adds an auxiliary header to the network payload.
5. Generates the MIC (if necessary).
6. The MIC is added to the frame, as shown in Figure 8.11.

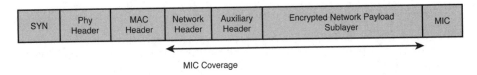

Figure 8.11 ZigBee frame format with network security

8.9 APPLICATION SUPPORT SUBLAYER SECURITY

The application support sublayer (APS) can apply security to outgoing and incoming frames using either the link or network key. It relies on the APS Information Base (AIB) for storing the following security attributes:

- **Application device key pair set:** The following are the elements of the key pair descriptor:

    Device address

    Master key (16 bytes)

    Link key (16 bytes)

    Outgoing frame counter (4 bytes)

    Incoming frame counter (4 bytes)

- **Application trust center address:** 64-bit address of trust center.

8.9.1 Applying Security to an Application Support SubLayer Outgoing Frame

Applying Security to a Frame Generated by APSDE-DATA.Request

This type of frame contains the Tx Option field, as shown in Figure 6.2 (in Chapter 6, "Zig-Bee Application Support Sublayer (APS)"). The Tx Option identifies which key type should be used for security. If Tx Option is 01, the link key is used; if the Tx Option is 02, the network key is used.

- Using the link key for security

  The APS uses the destination address to find a key pair descriptor. If a key pair descriptor is found, the APS uses the link key to apply security to the APS frame and sets the key identifier to 0x00 in the auxiliary header.

- If the Tx Option equals 0x2, the APS should use the network key for applying security. The security material can be obtained by using the nwkActiveKeySeqNumber from the NIB security attributes. The key identifier is then set to 01 in the auxiliary header.

Applying Security to an APS Command Frame

The APS queries the apsDeviceKeyPairSet using the destination address. If there is a key pair descriptor for the destination address, the APS uses the link key; otherwise; the network key is used.

8.9.2 Applying Security to an APS Incoming Frame

The auxiliary header of an incoming frame contains information about the security of the frame. Figure 8.12 shows the network layer auxiliary header.

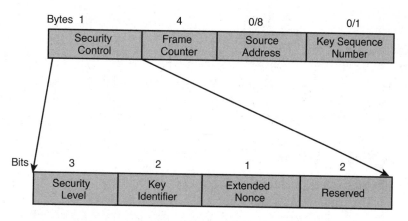

Figure 8.12   Network layer auxiliary header

If the security level is greater than 0, security has been applied to the frame and the key identifier determines the type of key (link or network) that was used. The key identifier can have following values.

- **Key identifier = 00: Link key (data key).** The security information (the link key and incoming frame counter) for the source device is defined by the application device key pair set attribute stored in the AIB.
- **Key identifier = 01: Network key.** The security information for the source is defined by the network security material attribute stored in the NIB.
- **Key identifier = 02: Transport key.** The security information is located in the application device pair set attribute and the transport key is derived by applying a keyed hash function to the link key with an input string of 0x00.
- **Key identifier = 03: Loaded key.** The security information is located in the application device pair set attribute, and the loaded key is derived by applying a keyed hash function to the link key with an input string of 0x02.

8.9.3 Transporting Keys

ZigBee offers unsecured transport of keys and secured transport of keys. In the secured transport of keys, the trust center handles the transport of the link, network, and master keys to the various devices. In the unsecured transport of keys, the devices are loaded with the keys.

The following methods can be used to install security keys in a ZigBee device.

- **Out of band:** The trust center transmits the key to a device using a different channel than the normal communication channels used by the network.
- **In band:** The trust center distributes the key to the devices using the normal communication channels. This method is less secure because a new device receives its key and then requests to join the network.
- **Preinstalled by manufacturer:** The manufacturer of the ZigBee device initially loads the key onto the device. That is, the manufacturer gives the key to the user of the device. This can be considered less secure because a third party, the manufacturer of the device, knows the key.

8.9.4 Key Establishment

The key establishment procedure is used to generate link keys for two devices using their master keys. The APS management entity (APSME) offers primitives for generating a link key for two devices. However, the two devices must have already installed a master key to use the SKKE protocol for generating the link key. Figure 8.13 shows the key establishment process.

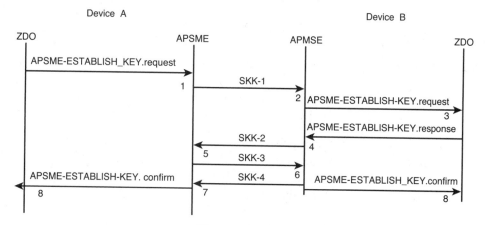

Figure 8.13   Generating a link key using a key establishment procedure

1. The ZigBee device object (ZDO) initiates key establishment by sending an APSME-ESTABLISH-KEY.Request to its local APSME, as shown in Figure 8.13. Figure 8.14 shows the APSME-ESTABLISH-KEY.Request primitive.

Responder Address	Use Parent	Responder Parent Address	Key Establishment

Figure 8.14   APSME-ESTABLISH-KEY.Request

- **Responder address:** Address of the responder.
- **Use parent (true/false):** True indicates that the APSME-ESTABLISH-KEY.Request should be sent to the device's parent and the parent would then forward the command to the device; otherwise, do not use the parent, and instead send the command to the device.
- **Responder parent address:** If the use Parent field is set to true, stores the address of the parent.
- **Key establishment method (0x00–0x03):** 0x00 uses the SKKE protocol, 0x01 to 0x03 are reserved.

2. The APSME executes the SKKE protocol and transmits the SKK-1 command frame to the responder's APSME. Figure 8.15 the shows general frame format of an SKKE command.
   - **Frame control:** Same as Figure 6.7
   - **Command identifier:** Defines the type of command, including the following:
     - 01 SKKE-1
     - 02 SKKE-2
     - 03 SKKE-3
     - 04 SKKE-4
     - 05 Transport-Key

Frame Control	Command Identifier	Initiator Address	Responder Address	Data

Figure 8.15   General frame format of SKKE command

- **Data:** The data value depends on the command type. For an SKKE-1 command frame, it is the challenge bit string called the QEU that is generated by the initiator.

3. When the APSME of the responder receives the SKKE-1, it informs the ZDO of the responder by sending an APSME-ESTABLISH-KEY.Indication primitive. The ZDO can either accept or reject the key establishment. Figure 8.16 shows the APSME-ESTABLISHMENT-KEY.Indication primitive.

Initiator Address	Key Establishment Method

Figure 8.16   APSME-ESTABLISHMENT-KEY.Indication primitive

- **Initiator address:** Initiator device address.
- **Key establishment method:** 00 means SKKE; 01, 02, and 03 are reserved.

4. The ZDO of the responder will respond by sending an APSME-ESTABLISH-KEY.Response to its local APSME, as shown in Figure 8.17.

Initiator Address	Accept

Figure 8.17   APSME-ESTABLISH-KEY.Response primitive

- **Accept (true/false):** If true, the responder's ZDO accepts the initiator's key establishment; otherwise, it is rejected. If the ZDO accepts the key establishment, the process continues through steps 5, 6, and 7. Otherwise, the process goes to step 7.

5. The responder's APSME sends an SKKE-2 to the initiator's APSME. The format of SKK-2 is same as Figure 8.15, where the Command Identifier field is 02 and the Data field is a challenge bit string generated by the responder called the QEV.

   When the initiator's APSME receives the SKKE-2, it transmits the SKKE-3 to the responder with the command identifier set to 03 and the Data field containing the MACtag2. The MACtag2 (link key) is the message authentication code (MAC) that is generated from the bit string of the MACData2. The MACData2 is represented by the following:

   MACData2 = (U ∥ V ∥ QEU ∥ QEV)

where

U is the identification of the initiator.

V is the identification of the responder.

QEU is the string of challenge bits generated by the initiator.

QEV is the string of challenge bits generated by the responder.

|| means concatenation of U, V, QEU, and QEV.

The Matyas-Meyer-Oseas method, using the master key, is applied to the MACData2 to generate the link key for the responder.

6. When the responder's APSME receives the SKKE-3, it sends an SKKE-4 to the initiator's APSME, where the Command Identifier field is set to 04, and the Data field is a bit string called MACtag1.

   As in step 5, the MACtag1 is the MAC that is generated from the bit string of MAC-Data1 and master key using the HMAC Matyas-Meyer-Oseas method. Similarly, the MACtag1 is the link key of the initiator.

7. The APSME of the initiator and responder each send an APSME-ESTABLISHMENT-KEY.Confirm to their respective ZDOs. Figure 8.18 shows APSME-ESTABLISH-MENT-KEY.Confirm primitive.

Figure 8.18   APSME-ESTABLISHMENT-KEY.Confirm primitive

- **Address:** Address of the device that executes the SKKE protocol

  Both the initiator and responder compare their MACtag1 and MACtag2; if they are equal, the message authentication code is the link key.

  Some status values and their descriptions are as follows:

00	Success
01	Invalid parameter
02	No master key
03	Invalid challenge
06	Invalid key

SUMMARY

This chapter addressed ZigBee security, including the following topics:

- Elements of network security are secrecy, authentication, integrity, and nonrepudiation.
- Secrecy provides privacy and protects information from being intercepted. If a person intercepts the information, the information would have no meaning.
- Authentication methods verify the identity of a person or computer accessing the network.
- Integrity maintains data consistency and prevents tampering with information.
- Nonrepudiation provides proof of origin to the recipient.
- In symmetric cryptography, the transmitter and receiver of a message share a key.
- In asymmetric cryptography, the transmitter and receiver of a message are using different keys.
- ZigBee can apply cryptography and frame integrity at the application and network layers.
- ZigBee security uses counter mode (CTR) with 128-bit AES for encryption of the message.
- ZigBee security uses cipher block chaining (CBC) with 128-bit AES for generating the message integrity code (MIC).
- ZigBee defines three different keys: the link key (secret key), network key, and master key.
- ZigBee defines two types of application key: application link key and trust center application link key.
- The application link key is used for securing application data between two devices.
- The application link key can be configured and distributed by the trust center or installed using the SKKE protocol.
- The application link key is shared only between two devices.
- The network key is used for security of broadcast messages and is shared by all devices.
- The master key is used to generate the link key, and it can be preinstalled by the manufacturer of the node or distributed by the trust center.
- ZigBee defines two security modes: standard and high.
- ZigBee uses a trust center (coordinator or dedicated device) for distributing keys to the devices in the network and authenticating new devices joining the network.
- ZigBee uses a sequential freshness counter to prevent replay attacks.
- Security keys can be transported out of band or in band or can be preinstalled by the manufacturer.

REFERENCES

1. Advanced Encryption Standard, www.csrc.nist.gov/publications

2. Ember Corp, Ember Znet Application Developer's Reference Manual, 2008

3. www.zigbee.org/imwp/idms/popups/pop_download.asp?contentID=9436, ZigBee Alliance Presentation

4. www.csrp.inl.gov/Documents/Securing%20ZigBee, Recommended Practices Guide for Securing ZigBee Wireless Networks

5. Document 053474r17, ZigBee Specification, ZigBee Alliance, October 2007

6. Freescale BeeStack Software Reference Manual, Freescale Semiconductor, July 2007

7. Elahi & Elahi, *Data, Network & Internet Communications Technology,* Thomson Learning, 2006

8. Yuksel, Nielson, ZigBee - 2007 Security Essential, Proceedings of the 13th Nordic Workshop on Secure IT Systems, NordSec 2008 , pages: 65-82, pages: 212, 2008, DTU-IMM, Denmark

CHAPTER 9

ADDRESS ASSIGNMENT AND ROUTING

INTRODUCTION

The function of the network layer is to assign a 16-bit address to each device joining the network and to route packets to their destination. ZigBee PRO uses a stochastic (nondeterministic) scheme to assign the address of each device in a network, whereas a network conforming to a tree topology uses a distributed scheme (Cskip) to assign a 16-bit address to each device.

9.1 ADDRESS ASSIGNMENT USING DISTRIBUTED SCHEME

When a device joins a network, it receives an address. This 16-bit address is assigned to the device by a router or coordinator and is unique within the current network. If the new device joining a network is a router, the coordinator assigns an address to it; otherwise, if the new device is an end device, either a router or the coordinator will assign it an address.

In a tree topology, ZigBee uses an algorithm called Cskip for assigning an address to each device. The following attributes, which are relevant to the Cskip algorithm, are stored in the Network Information Base (NIB) of the coordinator:

- Network maximum depth, represented in the Equation 9.1 as Lm
- Network maximum children for each router, represented as Cm
- Network maximum number of routers, represented as Rm

The ZigBee coordinator assigns a block of addresses to each router based on the maximum number of children. Each parent router uses the Cskip function to assign an address to each end device.

Figure 9.1 shows a network with a tree topology that consists of a coordinator, four routers (R1, R2, R3, and R4), and eight end devices. Assume each router can have maximum of three children.

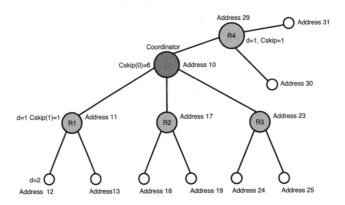

Figure 9.1   A network with tree topology

In Figure 9.1, $Lm = 2$, $Rm = 4$, $Cm = 3$, and the current depth represented by d. Equations 9.1 and 9.2 are used to find the value of $Cskip(d)$.

$$Cskip(d) = 1 + Cm(Lm - d) \quad if \ Rm = 1 \tag{9.1}$$

otherwise

$$Cskip = \frac{1 + Cm - Rm - Cm.Rm^{Lm-d-1}}{1 - Rm} \quad if \ Rm > 1 \tag{9.2}$$

At the coordinator, the value of the depth of the network d is 0. Therefore

$$Cskip(0) = 1 + 3 - 4 - 3 \times 4^{2-0-1} / \ 1 - 3 = 6$$

The address of the first router is equal to the address of the coordinator +1. For example, if the address of the coordinator is 10, the address of the first router $R1 = 10 + 1 = 11$.

The following equation is used to assign an address to the next router:

$$R_n = R_1 + (n - 1) * Cskip(d)$$

where R_n is the address of the router n

$$R_2 = \ 11 + Cskip(0) = 11 + 6 = 17$$
$$R_3 = \ 11 + 2*Cskip(0) = 23$$
$$R_4 = \ 11 + 3*Cskip(0) = 29$$

The following equation is used by the router to assign an address to each of its end devices.

The value of *Cskip(d)* for $d = 1$ is

$$Cskip(1) = 1 + 3 - 4 - 3 \times 4^{2-1-1}/1 - 3 = 1$$

Therefore, each router uses *Cskip* as an offset to assign addresses to their children.

Router R_1 initially assigns an address to one of its children, which is one greater than its own address. It then uses the *Cskip(1)* as the offset value to assign addresses to the other children connected to it. The same procedure is used by routers R_2, R_3, and R_4.

9.2 STOCHASTIC ADDRESS ASSIGNMENT

ZigBee PRO offers stochastic address assignment. This means that each node, when joining the network, is assigned a random number from 0 to 65,536. When a new device joins the network, it broadcasts a device announcement to inform the other devices of its presence in the network. If any device in the network discovers an address conflict with the new device, a conflict notification will be broadcasted and the parent of the child will assign another address to the child.

9.3 ROUTING

The function of the router in a ZigBee network is to provide a routing protocol so that a message may be routed from a source to a destination. ZigBee uses tree routing, ad hoc on-demand distance vector (AODV), many-to-one, and source routing.

9.3.1 Ad Hoc On–Demand Distance Vector (AODV) Routing Protocol

A ZigBee mesh network uses AODV for its routing protocol to determine the route from a source to a destination. The AODV dynamically establishes the route and can adapt to changes in the network. As its name implies, AODV is an on-demand routing protocol. When a source device does not have route to the destination, it initiates route discovery by broadcasting a route request (RREQ) command to its neighbors. It builds routes using route request and route reply commands. AODV is capable of both unicast and multicast routing.

9.3.2 Unicast Route Discovery

The unicast route discovery request establishes a route to one destination using the following process:

1. When a source node requires a route to a destination and does not have a route to the destination within its routing table, it broadcasts an RREQ packet across the network. The RREQ packet contains the following fields:

Source Address

Destination Address

Broadcast ID

2. Each RREQ packet is identified by its source address and broadcast ID. The node receiving the RREQ packet examines it to determine whether it has already received this RREQ from its other links, as shown in Figure 9.2. If a node has received several RREQ packets, it selects the one with the least routing cost and uses the information in the RREQ to make a reverse route entry in its routing table.

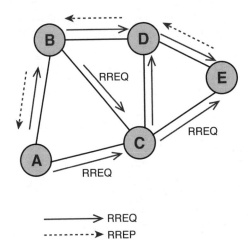

Figure 9.2 Route discovery process

3. When an intermediate node sets its reverse route, it checks its routing table for a route to the destination. If it has a route to the destination, it sends a route reply packet (RREP) to the source so that the source will be able to transfer the packet to the destination.

4. If an intermediate node does not have a route to the destination, it broadcasts an RREQ to its neighbors, and this process will continue until the RREQ packet reaches the destination.

5. When the RREQ packet reaches the destination, the destination node transmits a unicast RREP packet to the neighbor that sent the RREQ packet. This RREP packet

propagates back to the source through the intermediate nodes to set up the forward path to the destination.

6. When a source device receives the RREP packet, it can then transmit the data packet to the destination.

A route is considered active as long as data packets are periodically traveling from the source to the destination. If the link between B and D becomes broken for the active route A-B-D-E, as shown in Figure 9.3, node B detects the broken link and sends a route error (RERR) packet to source node. The source node will then reinitiate route discovery.

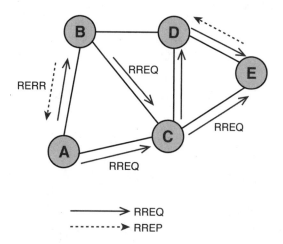

Figure 9.3   A network with a broken link

9.3.3 Multicast Route Discovery

Multicast route discovery is similar to unicast route discovery. If a node receives an RREQ that is not a member of the group or does not have a route to any group member, it creates a reverse route to the source and broadcasts an RREQ to its neighbors. When a node receives an RREQ and it is a member of the group, it sends an RREP packet to the source. The RREP packet propagation to the source sets up the forward path to the destination. Figure 9.4 shows the nodes D and E, which belong to a group, and node A, which requests route discovery by broadcasting an RREQ to its neighbors. Both D and E will send RREP packets to source A. The source caches the routes in its routing table.

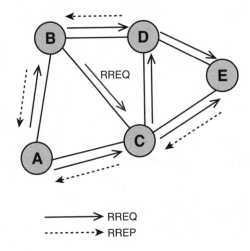

$$\longrightarrow \text{RREQ}$$
$$\text{-------} \rightarrow \text{RREP}$$

Figure 9.4 Multicast route discovery

9.4 DYNAMIC SOURCE ROUTING (DSR)

The Dynamic Source Routing protocol is a simple and efficient routing protocol that is designed for wireless ad hoc networks with multiple hops and wireless mesh networks. In a source routing protocol, the node holds in its route cache those addresses of the intermediate nodes that lie between the source and destination. When a source wants to transmit a packet to the destination, it adds the addresses of intermediate nodes to the packet and only then transmits the packet. DSR is a true source routing protocol: All routing information is maintained by the source node, and the intermediate nodes do not hold any information about the route. The protocol is composed of the two main mechanisms: route discovery and route maintenance. These modes operate together to allow nodes to discover and maintain routes to any destinations in an ad hoc network. DSR is an on-demand protocol, which means when a source does not have a route to the destination, it will request route discovery.

9.4.1 DSR Route Discovery

Route discovery is a mechanism that a source node uses to find a route to the destination when it does not already have the route in its route cache. Figure 9.5 shows a network with multiple hops in which source A requires a route to destination E. The following steps describe the route discovery process that is used for the network depicted in Figure 9.5.

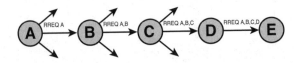

Figure 9.5 DSR route discovery

1. Node A transmits a broadcast RREQ to its neighbors (wireless nodes in its range). The RREQ packet has a field called Route Record that stores a list of intermediate node addresses.

2. Node B and other nodes in the range of node A receive the RREQ packet. Node B checks its route cache to determine whether it has a route to the destination or whether it is the destination; if it does (or is), it sends an RREP packet to A.

3. If the node that receives the RREQ is not the destination, it adds the source node address to the RREQ record and broadcasts an RREQ to its neighbors. In this example, B adds node A's address to an RREQ packet and transmits to its neighbor.

4. If node C is not the destination, it adds node B's address to the RREQ record and transmits to its neighbors.

5. Node D performs the same process, and eventually RREQ arrives at node E, the final destination.

6. Node E sends an RREP packet to node A using the reverse route in the RREQ packet.

7. Node A caches the route from the RREP packet and uses it for transmitting the packet to node E.

Each node keeps a copy of the RREQ packets recently received in its route request table. When a node receives an RREQ packet that exists in its route request table, it discards the packet so that it does not broadcast multiple RREQ packets.

9.4.2 Route Maintenance

In the DSR protocol, each node is responsible for packet delivery to the next node, as shown in Figure 9.5. For example, node A is responsible for the data packet to node B. This is done by receiving an acknowledgment packet from node B. Node B transmits the data packet to node C and requests an ACK frame. If node B does not receive an ACK frame from node C, node B will retry several times. If node B still does not receive an ACK frame, it considers the link between B and C to be broken. Node B removes node C from it route cache and sends a route error packet to the originator of the RREQ packet. The originator will remove all the routes that are using the link between B and C from its route cache.

9.5 ZIGBEE ROUTING

One of the functions of the ZigBee network layer is to route packets to their destinations. When the network layer receives a data packet from its upper layer, it performs the following tasks before packet delivery:

- If the device is a router or ZigBee coordinator, it receives a data packet for delivery from its upper layer. The network layer of the device checks the destination address to see whether the destination address is a child. If it is, it delivers the packet to the child.

- If the device is an end device, it searches its neighbor table for the destination address. If the address of the destination device is in the neighbor table, it delivers the packet to the destination.
- If the device has routing capability (meaning the device has a routing table and a route discovery table), it checks its routing table to find an entry for the destination address. If there is an entry with an active status, it will use the route to transmit the packet to the destination. If the entry has a status set to routing discovery in progress, the network layer will store the data packet in its buffer and wait for the route discovery to be completed. It will then use the route for transmitting the packet to the destination. If the device does not have a route to the destination, it transmits an RREQ.

Routing attributes: The following attributes are stored in the NIB:

- **Network discovery retry limit:** Determines how many times the network layer will try to issue the route discovery command; it is set to 3.
- **Network route discovery time:** It is set to 2,710 milliseconds.
- **Network maximum router:** Maximum number of routers in a personal-area network (PAN); it is between 0x01 and 0x0f.
- **Network uses tree address allocation (true/false):** If true, the coordinator uses distributed address allocation; otherwise, the application layer defines address allocation.
- **Network uses tree routing (true/false):** If true, the network layer uses hierarchical routing; otherwise, it does not.
- **Network next address:** When a device requests association, the coordinator or router assigns an address to the device and then increments the address using the network increment value.
- **Network available address:** The number of addresses available for assignment. Each time the coordinator or router assigns an address to a device, this number is decremented by one.
- **Network address increment:** Each time a router or the coordinator assigns an address to a device, the network address is incremented by the Cskip value.

9.5.1 Routing Cost

ZigBee uses a path cost metric for determining the best route to transfer a frame to the destination. Each link has a cost. A path from a source to the destination may go through several links, the cost of which is the sum of individual links. The ZigBee network layer uses Equation 9.3 to calculate the cost of each link.

$$Cost = Minimum\ value\ of\ (7,\ 1/p^4) \tag{9.3}$$

where p is the probability of the packet being delivered using the link.

The value of p is calculated during packet delivery. One method of calculating p is to use the average link quality indication (LQI). When a packet arrives at a node, its link network layer records the LQI of each packet. The average of several LQIs can be used as value for p. The LQI indicates the quality of the packet received by the receiver, and it can be measured by received signal strength (RSS) or by signal-to-noise ratio (SNR).

9.5.2 Tree Hieratical Routing

Consider Figure 9.1: The end device with a source address of 12 wants to transmit a frame to the end device with a destination address of 30.

The end device transmits the frame to router R1. Router R1 checks whether the address of destination is a descendent address by satisfying Equation 9.4.

$$A < D < A + Cskip(d-1) \tag{9.4}$$

where A is the source address and D is the destination address.

If $A < D < A + Cskip(d - 1)$ is true, the network layer routes the frame to the appropriate child; otherwise, it routes the frame to its parent.

By substitution, $A = 12$, $D = 1$, and $Cskip(1) = 1$ in Equation 9.4 results in $12 < 30 < 13$.

Through the equation, the router, R1, determines that the frame is not intended for any of its children. Router R1 will forward the frame to its parent (coordinator). The coordinator uses Equation 9.5 to find the address of the next router.

$$N = A + 1 + \left[\frac{D-(A+1)}{Cskip(d)} \right] * Cskip(d) \tag{9.5}$$

Note: When dividing $(D - (A + 1) / Cskip(d)$, select an integer value.

N is the address of the next router; A is the address of the current source (and in this case the coordinator address).

By substitution, $A = 10$, $D = 30$, $Cskip(0) = 6$ in Equation 9.5 results in the following:

$$N = 10 + 1 + [(30 - 11) / 6] \text{ x } 6 = 11 + [19 / 6] \text{ x } 6 = 11 + 3 \text{ x } 6 = 29$$

The next node is the router with an address of 29. The coordinator will then transmit the frame to the router with an address of 29 (R4). The router, R4, transmits the frame to designated address, 30.

The advantage of tree routing is that it is simple and uses limited resources. In tree routing, the router decides to forward the packet to its child or its parent. Therefore, the router does not require a significant amount of memory to store its routing table. The disadvantage of tree routing is that it is inefficient. If two neighboring nodes are connected to two different routers, they cannot exchange information directly. The packet exchange between nodes must travel through several routers to reach its ultimate destination.

9.5.3 Tree Routing Process

When a frame is transported from the upper layer to the network layer (for example, an NLDE-DATA.Request), the network layer performs the following functions for delivery of the frame:

- If the destination of the frame is set for broadcast, the frame is transmitted.
- If the device does not have routing capability, the NLME checks the "network uses tree routing" value from the NIB.
  1. If the "network uses tree routing" value is true, it uses tree routing for forwarding the frame to the destination.
  2. If the "network uses tree routing" value is false, there is no route to the destination, and the network layer discards the frame.
- If the device has routing capability, the network layer performs the following steps to route the frame to the destination:
  1. The network layer checks the value for Route Discovery field. If this field is set to 0x02 (force route discovery), the router should start route discovery.
  2. If the Route Discovery field is set to 0x01 and there is no routing entry for the destination in the routing table of the device, the router should start route discovery.
  3. If the Route Discovery field is set to 0x00 or 0x01, the network layer checks the routing table to find an entry by comparing the destination address in the frame and the destination address of routing table. If there is a match, it checks the status field in the routing table.
     a. If the Status field of the routing table is set to active, the network layer will forward the frame to the next hop.
     b. If the Status field of the routing table is set to discovery under way, the network layer may store the frame in its buffer or use tree routing if the "network uses tree route" value is set to true.
  4. If the device does not have routing table and route discovery is set to 0x00, the network layer checks network use tree route.
     a. If "network uses tree routing" is set to true, the network layer uses tree routing.
     b. If "network uses tree routing" is set to false, the frame is discarded.

9.5.4 ZigBee PRO Routing

ZigBee PRO offers aggregated routing by using many-to-one and source routing. In most of the wireless sensor networks, all nodes in the network communicate with one node called the concentrator or gateway. In building automation, all sensors controlling smoke detectors will transmit the status of the smoke detector to a gateway or concentrator. Figure 9.6 shows a network with a gateway (concentrator).

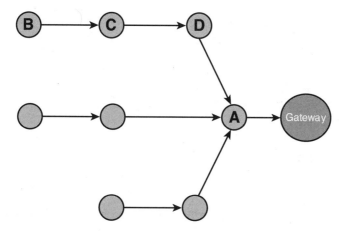

Figure 9.6   A network with a concentrator

As shown in Figure 9.6, if each node requests a route to the gateway, node A will over-flow with RREQ packets from all the other nodes. In source routing, the source node speci-fies the route to the destination. Assume node B routes a packet through nodes C, D, and A in order to reach the gateway. Node A needs to know only the address of node B to send packets to the gateway. However, in source routing, each device must hold the address of all devices in the route path. This requires that each node store a large routing table. In many-to-one routing with source routing, this requirement is eliminated. That is, the device is not required to store a large routing table. In many-to-one routing, each node is required to hold only the address of its neighbor and the gateway. The gateway uses source routing to trans-mit the data packet to the nodes. The gateway can support this because it generally has a larger memory than the other nodes in the network.

9.6 ZIGBEE ROUTING COMMANDS

Figure 9.7 shows the general network command frame, which includes the network Com-mand Identifier field. This field identifies the type of command. Possible values are listed here.

Command Identifier Value	Command Name
0x01	Route request
0x02	Route reply
0x03	Route errors
0x04	Leave
0x05	Route record
0x06	Rejoin request

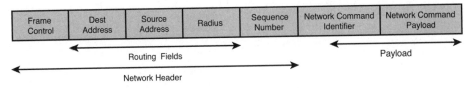

Figure 9.7 General command format

In Figure 9.7, the payload contains a network command identifier and information about the route.

9.6.1 Route Request

The route request is generated by a device to request a route to the destination. Figure 9.8 shows the frame format of the route request. The router request is a broadcast frame. Therefore, the destination address of network header should be set to 0xfffd. This means any device with MAC receiver on while idle set to true receives this packet and should hold a copy of the broadcast packet in its broadcast transaction record (BTR).

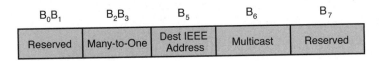

Figure 9.8 Route request frame format

The Command Option field is 8 bits, as shown in Figure 9.9.

Figure 9.9 Command option bits

B₂B₃

00	Route request is not many-to-one.
01	Route request is many-to-one and the transmitter supports route record.
10	Route request is many-to-one, and the transmitter does not support route record.

B₅

0	Command frame contains the network address of the destination.
1	Command frame contains the destination IEEE address.

B₆

1	Command frame is multicast, and the destination address holds the group.

- **Command identifier:** 01 route request
- **Route request identifier (8 bits):** This is a sequence number; it will be incremented by one each time the source issues a new route request.
- **Destination address (16 bits):** Address of the destination device.
- **Path cost:** The accumulated cost of the links.

9.6.2 Route Reply Command

This is generated in response to a router request. A device that receives an RREQ from the source will indicate the reception of the RREQ. Figure 9.10 shows the route reply command format.

Network Header	Network Command Identifier	Command Options	Route Request Identifier	Originated Address	Responder Address	Path Cost

Figure 9.10   Route reply frame format

- **Network identifier**: 02 router reply
- **Command option:** Indicates whether a repair of the route is required; bits 0 to 6 are reserved, and b7 is set to one for route repair.
- **Route request identifier:** This is the same value of the RREQ sequence number.
- **Originator address:** This is the 16-bit network address of the device that generated the route request.
- **Responder address:** Same as the destination address of Figure 9.8.
- **Path cost:** Defines the quality of the link.

9.6.3 Route Error

This command is used to notify the sender when a device is unable to forward the data frame to the destination based on the route that it was using. Figure 9.11 shows the route error command frame format.

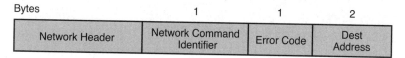

Bytes		1	1	2
Network Header	Network Command Identifier	Error Code	Dest Address	

Figure 9.11   Route error frame format

- **Network identifier:** 03 for route error.

  The ZigBee router is able to use tree routing by default or the AODV algorithm.

- **Error codes:**

  00 No route available.

  01 Tree link failure; this used if the router uses a tree route.

  02 Nontree link failure; this field will be used if the router uses AODV.

  03 Low battery.

  04 No routing capacity.

9.6.4 Routing Table

The ZigBee coordinator or router maintains a routing table. The elements of a routing table are as follows:

- **Destination address:** Address of final destination (16 bits).
- **Status:** Status defines the status of routing:

  0x00 Active.

  0x01 Route discovery under way.

  0x02 Inactive.

  0x03 Checking route validation.

- **No route cache:** This bit indicates that the destination does not hold the source route.
- **Many-to-one:** This bit is set to 1 when the destination is the concentrator.
- **Group ID:** Indicates the destination address is a group ID.
- **Next-hop address:** The address for the next hop.

Any device that maintains a routing table should have a routing table capacity. The ZigBee coordinator and router maintain a routing table with routing table capacity. The following list shows the elements of the routing table capacity:

- **Route request ID:** Sequence number of the route request
- **Source address:** 16-bit address of the source device that initiates the RREQ
- **Sender address:** Address of the sender device (16 bits)
- **Forward cost:** The total cost from the source to the current device
- **Residual cost:** The total cost from the current device to the destination

SUMMARY

- When a device joins a network, it acquires a 16-bit address.
- The network layer assigns the 16-bit address, which is unique within the network.
- Tree topology uses Cskip to assign the 16-bit address to each device in the network.
- ZigBee PRO uses a stochastic scheme to assign addresses to each device in the network.
- A ZigBee router offers tree routing, Dynamic Source Routing (DSR), and ad hoc on-demand distance vector (AODV) routing protocols.
- AODV offers unicast and multicast route discovery.
- AODV uses route request (RREQ) and route reply (RREP) commands for route discovery.
- In AOVD, each RREQ packet is identified by the source address and the broadcast ID.
- In a tree topology network, the network uses tree routing.
- In DSR, the packet carries the entire route to the destination.
- DSR is an on-demand routing protocol.
- ZigBee PRO uses a many-to-one source routing algorithm.
- The benefit of tree routing is that it is simple and uses limited resources.
- Tree routing is an efficient method for routing the packet to the destination.
- The ZigBee network layer uses route request, route reply, and route error commands.

REFERENCES

1. Deepankar Medhi, Karthikeyan Ramasamy, *Network Routing Algorithm, Protocols, and Architecture,* Morgan Kaufman, 2007

2. Tony Kenyon, *Data Networks Routing, Security, and Performance Optimization,* Digital Press, 2002

3. Document 053474r17, ZigBee Specification, ZigBee Alliance, October 2007

4. Drew Gibson, ZigBee Wireless Networking, Newnes, 2008

CHAPTER 10

ZigBee Home Automation and Smart Energy Network

10.1 HOME AUTOMATION PROFILE

The ZigBee Alliance developed the home automation (HA) profile, which defines a set of standards for the home automation network. Manufacturers of HA devices use these standards for interoperability between devices. The HA network is used for controlling lights, blinds, thermostats, home theater devices, occupancy detection, motion sensors, security systems, and door and window sensors in real time. The ZigBee HA profile defines the following specifications for home automation:

- **HA clusters**
- **HA network requirements**
- **Device descriptions:**
  Device function
  Device supported cluster
- **Commissioning**

10.1.1 Home Automation Clusters

The HA clusters, listed here, are categorized based on their function:

- **General clusters:** Common to all ZigBee Alliance profiles.
- **Measuring and sensing clusters**: Clusters that are used for measurement and sensing; some examples are luminance measurement, pressure measurement, temperature measurement, flow measurement, occupancy sensing, and luminance-level sensing.

- **HVAC clusters:** HVAC clusters are used for heating and cooling control (for example, a thermostat) and fan control and configuration.
- **Security and safety:** Intruder alarm system device.

10.1.2 Home Automation Network Requirements

The ZigBee HA profile describes the following requirements and recommendations for an HA network:

- Device polling rate is 7.5 second, except for commissioning, which can be higher.
- HA does not use a concentrator; it does not support many-to-one routing, and the routing table can be increased as much as possible. When a new device joins the network, the coordinator is required to inform the installer of the network.
- The HA profile recommends channels 11, 14, 15, 19, 20, 24, and 25 as the operating channels.
- The HA profile does not recommend use of broadcast addressing.
- HA uses standard security.
- ZigBee HA defines the following startup attributes for interoperability between different manufacturers of ZigBee devices:

  Short address set to 0xFFF

  PAN ID set to 0

  Channel mask: All channels in frequency band

  Protocol version 2006 or higher

  Trust center address set to 0

  Network key set to null

  Master key set to null for the 2007 protocol stack

  Number of scan attempts to find a coordinator set to 3

  Rejoin interval set to 60 seconds and maximum interval set to 15 minutes (If a device fails to rejoin a network, it has to wait 15 minutes before starting the rejoin process.)

  Concentrator radius set to 5

  End binding timeout set to 60 seconds

  Support return to factory defaults

  Support frequency agility using the ZigBee 2007 protocol

10.1.3 Devices in Home Automation

The devices used in HA are categorized as generic, lighting, closures, HVAC, or intruder alarm systems. Table 10.1 lists devices used for HA with their corresponding device ID. The HA network can have between 2 to 500 nodes using the ZigBee feature set or Pro devices.

Table 10.1   Home Automation Devices and Corresponding Device ID

Device Name	Device ID
Generic	
On/off switch	0x0000
Level control switch	0x0001
On/off output	0x0002
Level control output	0x0003
Scene selector	0x0004
Configuration tool	0x0005
Remote control	0x0006
Lighting	
On/off light	0x0100
Dimmable light	0x0101
On/off light switch	0x0103
Dimmer switch	0x0105
Light sensor	0x0106
Occupancy sensor	0x0107
Closure	
Shade	0x0200
Shade controller	0x0201
HVAC	
Heating and cooling unit	0x0300
Thermostat	0x0301
Temperature sensor	0x0302
Pump	0x0303
Pump controller	0x0304
Pressure sensor	0x0305
Flow sensor	0x0306
Intruder Alarm System	
IAS control / indicating equipment	0x0400
IAS ancillary control equipment	0x0401
IAS warning device	0x0403

The ZigBee HA profile defines functions that are supported by each device. Table 10.2 shows the mandatory functions supported by an on/off switch.

Table 10.2 Functions Supported by an On/Off Switch

Device Function	Device Type
Join	Router and end device
Forming network	Coordinator only
Allow other devices to join network	Router and coordinator
End device bind request	End device
Binding response	Coordinator
Group node	Device
Service discovery request (matched description)	Device
ZigBee device profile bind response	Device
ZigBee device profile unbinding	Device
Service discovery request (matched description)	Device

10.1.4 Commissioning

ZigBee commissioning is a method to configure devices in a ZigBee network. Commissioning is a tool that enables the installer to install devices, check network operation, and troubleshoot. This tool can be a laptop, a PDA, or a simple push-button switch on a ZigBee device. The ZigBee Alliance defines a commission cluster that transmits commands over the air to the device. The function of the ZigBee commission tool depends on the type of ZigBee device and the applications that are used. In general, and besides being easy to use, the commission should be able to

- Configure the device's startup attributes (join a network, whether the device is a coordinator, form a network, and so on)
- Configure which devices have permission to join a network (membership)
- Discover the network and devices in the network
- Set up binding between devices
- Set up group membership
- Update the ZigBee protocol stack over the air
- Analyze the network and identify a problem if one occurs
- Recover device information, such as the following:

    Addresses (network and IEEE address).

    Link quality

    Manufacturer

- Analyze network performance
- Make changes in device configuration

The ZigBee Alliance defines three types of commissioning modes:

- **Automatic mode (A-mode):** Plug and play; a device automatically configures itself.
- **Easy mode (E-mode):** Devices contain some switches that the user sets to configure the device.
- **System mode (S-mode):** A laptop or PDA device is used to install the device.

The HA profile recommends E-mode for commissioning. In this mode, there are two push button switches, red and green, on each device. The following list describes the actions that occur when the red or green button is pressed:

- **Join network:** If you press the red button four times, the end node sends a request to join the network.
- **Form a Network:** If you press the green button four times, the device is "told" to form a network using the network-formation process.
- **Allow other devices to join the network:** If you press and hold down the red button and press the green button four times, you enable the router or coordinator to accept new devices.
- **Restore factory default of the device:** Holding the red and green buttons simultaneously for 60 seconds will restore the factory defaults.
- **End device binding:** Press the red button five times and then the green button five times on each device for binding. You do this to bind two devices together, (for example, a switch and light). Each device will send a bind request to the coordinator, and the coordinator will act as the binding manager.
- **Enable identity mode:** Press the red button six times and then the green button six times to place the device in identify mode for 60 seconds.

10.2 SMART ENERGY NETWORK

A smart energy network is a combination of an advanced metering infrastructure (AMI) and home-area network (HAN), as shown in Figure 10.1. AMI technology is used to read electric, gas, and water meters remotely; it can collect energy-usage information at any time from customers.

The smart energy network uses two-way communication between the HAN and the AMI. The neighbor-area network (NAN) is connected to the HAN by the energy service portal (ESP); the ESP typically can be a meter or a gateway.

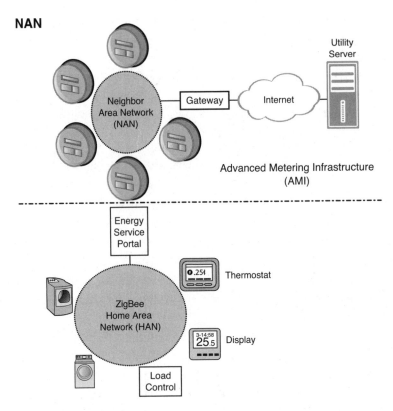

Figure 10.1 Smart energy network

10.2.1 Application of a Smart Energy Network

AMI enables the utility companies to have near-real-time information about electricity consumption. This data can then be used to set up a peak time for each part of the city or county, allowing the utility companies to better manage the distribution of electricity. The smart energy network can perform the following functions.

- Communicates in two ways (from meter to central utility and from central utility to meter).
- The utility company can read the meters at any time.
- Smart energy enables the utility companies to give specific information to customers regarding how to save energy.
- Reports power interruption and location.
- The meter can receive commands and, in the case of emergency, turn off customer power.

- Notifies customer of high peak, allowing the customer to reduce consumption.
- Customers can receive time-based pricing, allowing them to make smart energy choices.
- Time-based billing report (monthly, weekly, or daily).

10.2.2 Advanced Metering Infrastructure (AMI)

The AMI connects a NAN to the utility network. The NAN is a mesh topology network that connects thousands of smart meters to the utility application server through a public network.

- **Energy service portal (ESP):** Connects the energy supply network to the HAN. The ESP can be installed in a meter or separate device and supports the following clusters.
  - Energy price
  - Time
  - Load control
  - Metering
  - Prepayments

10.2.3 Home–Area Network (HAN)

The HAN is a wireless connection of thermostats, lighting systems, appliances, and display devices. The display devices display the energy consumption of the customer; they can display the current energy usage, price information, and energy-usage history (day by day).

The HAN enables the homeowner to manage consumption of electricity by using smart appliances and thermostats. The thermostat can be programmed for different temperatures at different times. By interfacing the NAN with the HAN, the utility company can provide real-time information about electricity prices to customers. The customer can decide when to turn on appliances. The following are typical smart energy devices used in a HAN:

- **Metering devices:** Electricity, gas, water, and submetering devices.
- **In-home display:** Used to display energy consumption, pricing, load profile, and informs customers of potential energy savings.
- **Programmable communication thermostats (PTC):** The programmable thermostat is used to control heating and cooling of the house.
- **Load controller:** The load controller is used to postpone the load from peak time to other times. This provides the customer a financial benefit by reducing usage during peak consumption times. The load controller receives commands from the ESP to control the consumption of energy by communicating with smart appliances and the heater.

10.3 ZIGBEE STACK PROFILE FOR SMART ENERGY (SE) PROFILE

The ZigBee Alliance published the ZigBee smart energy profile in May 2008. This profile defines a set of specifications for interoperability in a HAN between devices from different vendors. These specifications include the following:

- Startup attributes
- HAN security
- Specify devices with their ID used for HAN
- Specify smart energy clusters
- Specify inter-PAN communication
- Commissioning

10.3.1 Startup Attributes

For interoperability between ZigBee manufacturers of smart devices, ZigBee defines startup attributes that can be set by the installer of the ZigBee device or preprogrammed in the device:

- Device short address set to 0xFFFF.
- EPAN ID set to 0 (64 bits).
- PAN ID set to 0xFFFF.
- Stack profile set to 1 or 2.
- Trust center address set to 0x0000.
- The smart energy profile uses standard security and does not require a master key.
- Scan attempts set to 3. A device that wants to join a network scans all channels three times to find a ZigBee coordinator.
- Time between scanning channels set 1 second.
- Rejoin interval set to 60 seconds. A device can rejoin the network after 60 seconds.
- Maximum rejoin interval set to 15 minutes. If a device's rejoin attempt fails, it must wait 15 minutes to rejoin.
- Trust center transmits a network key to the devices in the network.
- The smart energy profile does not recommend broadcast addressing, except for the pricing cluster.

- The smart energy profile recommends frequency agility and frame segmentation, and the device polling rate should not be more than every 7.5 seconds, except during commission, when polling can be a higher rate.

10.3.2 Home Area Network Security

One of the important factors for successful deployment of a smart energy system is security. The smart energy profile recommends use of link and network keys for security. The ESP acts as trust center in a HAN. The trust center performs the following functions:

- If a device rejoins the network using a secure method, the trust center does not take any action.
- If a device rejoins the network using an unsecured method, the trust center sends an AP command to the parent of the device to remove this device from its children table.
- If the trust center finds that the device joined the network through key establishment, the trust center should send the device an updated network key.
- The trust center periodically updates the network key and link key.
- The trust center keeps track of devices using a preinstalled link key and a certificate based on the key establishment link key.

10.3.3 Joining a Smart Energy Network

A device attempting to join the network requires a trusted link key for the trust center to authenticate it. The trust center link key is installed on a device using an out-of-band preconfigured link key process.

Preconfigured trust center link key: The following steps describe a preconfigured trust center link:

1. The manufacturer of a device installs a code on the device. This code can be 48, 64, or 96, or 128 bits.
2. The CRC-16 $X^{16} + X^{12} + X^5 + 1$ is used to find the FCS of the manufacturer installed code.
3. The FCS is appended to the manufacturer code.
4. The link key is generated from the installed code with the appended FCS by using the one-way compression method, Matyas-Meyer-Oseas.
5. The installed code in the device is sent out of band to the trust center, and the trust center uses the same procedure to derive the link key. When the trust center has the same link key as the device, it can authenticate the device.

When the device joins the network, the trust center sends the network key to the device by using the key-transport key. Within the key-transport key, the network key is encrypted by the preinstalled trust center link key and transported to the device.

The trust center cannot use the preinstalled link key to send application data. The trust center and device require a new link; the device uses the key establishment method to generate a new link key.

10.3.4 Key Establishment

There are two types of key establishment:

- **Symmetric key establishment:** Both devices use the same key.
- **Public key establishment:** Smart energy uses implicit certificates for public key establishment. The implicit certificate is signed by the certificate authority (CA) and provides binding authentication between two devices. This process is known as certificate-based key establishment (CBKE). The ESP can act as a CA and transmit the certificate to the devices in the network.

After a node has joined the network and is authenticated by the trust center, the device initiates key establishment to obtain a new link key. When a node receives an updated link key, it can then communicate with the trust center. If a node needs to communicate with another node, both nodes require a link key, which is used in the following process of node communication:

1. The source node requests binding from the destination node.
2. The destination node responds to the bind request.
3. The source node requests a link key from the trust center for itself and the destination node by sending destination information to the ESP.
4. The ESP transports the link key to the source node and destination node, allowing the two nodes to communicate.

Communication between a device and the trust center requires a secure link key. One way to acquire a secure link key is by using implicit certificates. The implicit certificate is generated by a CA.

10.3.5 Implicit Certificates

Implicit certificates use elliptic curve cryptography (ECC) and MQV (Menezes-Qu-Vanstone) for its authentication protocol. Implicit certificates are similar to X.509 certificates but are smaller in code and faster. Explicit certificates are composed of three items: device

ID, public key, and digital signature. The Certicon Corp developed a powerful encryption method using modified ECC for implicit certificates.

In conventional certificates, the public key is part of the certificate, but in implicit certificates, the user is required to derive the public key from the certificate.

10.4 SMART ENERGY CLUSTER

The smart energy cluster is organized into two categories: general and smart energy. Table 10.3 shows the smart energy cluster and the corresponding cluster ID and security key.

Table 10.3   Smart Energy Cluster and Corresponding Cluster ID with Security Key

Cluster Type	Cluster Name	Cluster ID	Security Key
General	Basic	0x0000	Network
General	Identify	0x0003	Network
General	Alarms	0x0009	Network
General	Time	0x000A	Application link
General	Commissioning	0x0015	Application link
General	Power configuration	0x0001	Network
General	Key establishment	0x0800	Network
Smart energy	Price	0x0700	Application link
Smart energy	Demand response and load control	0x0701	Application link
Smart energy	Simple metering	0x0702	Application link
Smart energy	Message	0x0703	Application link
Smart energy	Complex metering	0x0704	Application link
Smart energy	Prepayment	0x0705	Application link

10.5 SMART ENERGY DEVICE

Table 10.4 shows devices used in the smart energy profile with corresponding device ID.

Table 10.4   Devices in the Smart Energy Profile with the Corresponding Device ID

Device Name	Device ID
Range extender	0x0008
Energy service portal	0x0500
Metering device	0x0501
Display	0x0502
Programmable communicating thermostat	0x0503
Load device control	0x0504
Smart appliance	0x0505
Prepayment terminal	0x0506

Energy service portal: The service portal interfaces the AMI to the HAN for energy management and meter reading. This may be a stand-alone device or installed in a meter, as shown in Figure 10.1. The ESP acts as a server and supports message, price, and demand response and load control clusters

Inter-PAN communication: The ZigBee Alliance adds a special "stub" to the ZigBee protocol stack for handling unsecured communications between two ZigBee PANs. This enables the pricing information to be transmitted to a device (display) located in another PAN.

SUMMARY

- ZigBee PRO is used in home-area networks (HANs).
- A HAN is a wireless connection of thermostats, lighting systems, appliances, and displays.
- A neighbor-area network (NAN) is a wireless connection of smart meters.
- Advanced metering infrastructure (AMI) is a connection of NANs and of advanced meters to the utility server.
- The energy service portal (ESP) is used to connect the energy supply network to a HAN.
- The ESP informs the customer of electricity pricing, peak time, and daily usage.
- The ZigBee Alliance developed the smart energy profile for interoperability between devices made by different manufacturers.
- The ZigBee smart energy profile defines security, device type, device cluster, and device ID.
- The application key, fragmentation, and polling rate of 7.5 seconds are mandatory for the ZigBee energy profile.
- ZigBee smart energy defines startup attributes for ZigBee devices.
- ZigBee smart energy defines methods for a device joining a network.
- ZigBee smart energy profile uses a public key for security.
- ZigBee smart energy uses implicit certificate for security.
- ZigBee smart energy offers commissioning for installing ZigBee devices.
- There are three commissioning methods: automatic mode (A-mode), easy mode (E-mode) and system mode (S-mode).

REFERENCES

1. www.ferc.gov/EventCalendar/Files/20070423091846-EPRI%20-%20Advanced%20Metering.pdf

2. www.eei.org/ourissues/electricitydistribution/Documents/quantifying_benefits_final_append.pdf

3. www.dps.state.ny.us/NYSEG_RGE_AMI_Filing.pdf

4. www.eds.com/services/solutionoverview/advancedmeter/

5. ZigBee Alliance, ZigBee Document 053474r17, ZigBee Specification

6. ZigBee Alliance, ZigBee Document 053520r25, ZigBee Home Automation Profile Specification

7. ZigBee Alliance, ZigBee Document 075123r01, ZigBee Cluster Library Specification

8. "Understanding ZigBee Commissioning," www.daintree.net/resources/whitepaper

CHAPTER 11

ZIGBEE RF4CE

INTRODUCTION

Radio Frequency for Consumer Electronics (RF4CE) is a protocol developed by a consortium that includes companies such as Freescale, Texas Instruments, OKI, Panasonic, Philips, Samsung, and Sony. It defines a standard specification for designing remote-control devices for the TV, VCR, and DVD. The RF4CE consortium merged with ZigBee to produce the ZigBee RF4CE standard. Whereas most remote controls currently are based on infrared (IR) technology that requires line of sight, RF4CE does not have that limitation. The following are characteristics of an RF4CE device:

- Does not require line of sight to the receiver
- Supports two-way radio frequency (RF) communication between the controller node and the target device
- Communication between target devices
- Enables the remote control to display device status
- Supports paging to locate the remote control (by pressing a button on TV that causes the remote control to beep, making it easier to locate)
- Operates in 2.4GHz
- Supports frequency agility
- Supports power-saving mode
- Supports pairing scheme
- Uses a multistar topology

11.1 RF4CE NODES AND TOPOLOGY

RF4CE defines two types of nodes:

- **Controller node:** The controller node (the remote control) contains a subset of the RF4CE protocol and does not have capability to start a network. The controller node receives a personal-area network (PAN) ID and a short address from the target node by using the network discovery command. It then uses a pairing process to join the network.
- **Target node:** The target node contains the complete RF4CE protocol. It is used in consumer electronics such as TVs and in CD and DVD players. When the target node turns on, it starts a network as a coordinator, and then the controller node and target node can join the network.

RF4CE uses a multistar topology. In RF4CE, each target device with its remote control comprises a PAN. As shown in Figure 11.1, the CD player with the CD remote control (RC) make PAN1, the DVD player with its remote control make PAN2, and the VCR with its remote control make PAN3. These three PANs together comprise the remote-control network. The multifunction remote controller can control all three devices.

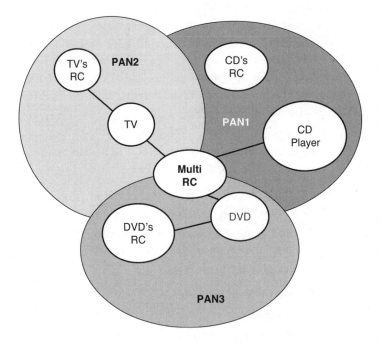

Figure 11.1 RF4CE network topology

11.2 RF4CE PROTOCOL ARCHITECTURE

Figure 11.2 shows the RF4CE protocol architecture, which consists of the IEEE 802.15.4 physical layer, Media Access Control (MAC) layer, network layer (developed by RF4CE), and application layer.

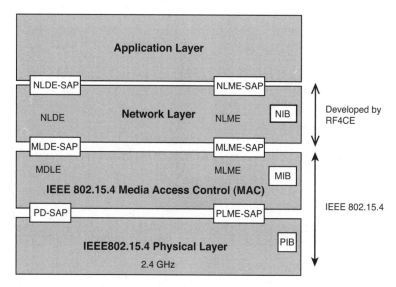

Figure 11.2   RF4CE protocol architecture

- **Physical layer:** The RF4CE physical layer uses the IEEE 802.15.4 physical layer with the following characteristics:

  Maximum physical layer payload is 127 bytes.

  Operates in the 2.4GHz band.

  Operational channels are 15, 20, and 25.

- **MAC layer:** RF4CE uses the IEEE 802.15.4 MAC layer and operates in a nonbeacon-enabled mode.

- **Network layer:** The RF4CE network layer provides management services through the network layer management entity (NLME) and data services through the network layer data service entity (NLDE). The network layer performs the following functions:

  Provides frequency agility to avoid interference

  Provides power-saving mode

  Supports the acknowledgment and unacknowledgment data transmission

  Supports unicast, multicast, and broadcast addressing

Starts a network (target node)

Pairing/Unpairing of two nodes

Discovery of nodes

11.3 NETWORK LAYER DATA SERVICES

The network data services are used for transmission of data between devices. The network layer provides the following data services:

- NLDE-DATA.Request
- NLDE-DATA.Response
- NLDE-DATA.Indication

The NLDE-DATA.Request is generated by the application layer and sent to the network layer for transmission of data. Figure 11.3 shows the format of the NLDE-DATA.Request primitive.

Device ID	Profile ID	Vendor ID	NSDU Length	NSDU	TX Option

Figure 11.3   NLDE-DATA.Request format

- **Device ID:** This is the reference to the pairing table.
- **Profile ID:** The profile ID that the device supports.
- **Vendor ID:** 0x0000, no vendor, or a valid vendor ID.
- **NSDU length:** Represents the number of bytes in the NSDU.
- **TX option:** 8-bit number: $B_7B_6B_4B_3B_2B_1B_0$.

  Where

  B_0 represents source address mode: $B_0 = 1$ means broadcast transmission, $B_0 = 0$ means unicast transmission.

  B_1 represents destination address mode: $B_1 = 1$ means broadcast transmission, $B_1 = 0$ means unicast.

  B_2 represents acknowledgment mode: $B_2 = 1$ means the source requests acknowledgment from the destination, $B_2 = 0$ means the source does not request acknowledgment from the destination. When the data process is completed, the destination will send a data confirm if the acknowledgment bit in the tx option is set to 1.

  B_3 represents security mode: $B_3 = 1$ means the packet is transmitted with security, $B_3 = 0$ means the packet is transmitted without security.

B_4 represents frequency agility: $B_4 = 1$ means frequency agility is enabled, $B_4 = 0$ means frequency agility is disabled and it uses only one channel for operation.

B_5 represents channel designation: $B_5 = 1$ means the operational channel is specified, $B_5 = 0$ means it does not specify the operation channel.

B_6 represents payload mode: $B_6 = 1$ means data in the payload is specified by the vendor, $B_6 = 0$ means data in the payload is not specified by the vendor.

B_7 Reserved.

- **NLDE-DATA.Confirm:** This primitive is generated by the destination to inform the source node of the status of NLDE-DATA.Request. It contains the following fields

  Status: Indicates the status of the NLDE-DATA.Request (for example, success or failed)

  Device ID: Indicates the reference to the pairing table

- **NLDE-DATA.Indication:** When the network layer of the destination receives the NLDE-DATA.Request, it sends the NLDE-DATA.Indication to its application layer.

11.4 NETWORK LAYER MANAGEMENT SERVICES

The network layer provides the following primitives for network management:

- NLME-START.Request
- NLME-AUTO-Discovery
- NLME-DISCOVERY.Request
- NLME-PAIR.Request
- NLME-PAIR.Response
- NLME-UNPAIR.Request
- NLME-UNPAIR.Response
- NLME-GET.Request
- NLME-SET.Request
- NLME-RXEnable.Request
- NLME-RESET.Request
- NLME-DISCOVERY.Response

The network layer can execute one process at a time; the processes are divided into two types, noninterruptible processes and interruptible processes:

- **Noninterruptible processes:** While RF4CE is processing noninterruptible processes, it cannot be interrupted by other processes. If RF4CE receives a new process while executing a noninterruptible process, it will ignore it. The noninterruptible processes are as follows:

  NLME-START.Request

  NLME-DISCOVERY.Response

  NLME-PAIR.Request

  NLME-PAIR.Response

  NLME-UNPAIR.Request

  NLME-UNPAIR.Response

- **Interruptible processes:** While RF4CE is executing any of the following processes, the RF4CE will be interrupted when receiving a new request. The following are interruptible processes:

  NLME-DISCOVERY.Request

  NLDE-DATA.Request

  NLME-AUTO-Discovery

  NLME-GET.Request

  NLME-SET.Request

  NLME-RXEnable.Request

  NLME-RESET.Request

11.5 NETWORK LAYER INFORMATION BASE (NIB)

The Network Information Base (NIB) is located in the network layer and contains attributes for the management of the network layer. These attributes include the following:

- **Network active period:** Determines the active period of a device.
- **Network base channel:** Selected channel during initialization.
- **Network discovery link quality indication (LQI) threshold:** The network discovery request will be rejected if its LQI value is below the LQI threshold.
- **Network discover repetition interval:** 1,000 MAC symbols.
- **Network duty cycle:** A value of 0x0000 indicates that the device does not use a duty cycle.
- **Network frame counter:** The frame counter added to the NPDU for replay protection.
- **Network indication discovery request (true or false):** True means the network layer will send a discovery indication primitive to inform the application layer of a discovery request; false means it will not.

- **Network power save (true/false):** True means the node operates in power-save mode; false means it does not.
- **Network pairing table:** Table that holds information about paring devices.
- **Network maximum discovery repetition:** Set to 16.
- **Network maximum number of backoff:** The maximum number of backoffs; it is used for carrier-sense multiple access with collision avoidance (CSMA/CA) before announcing a channel access failure.
- **Network maximum attempts frame retries**: The maximum number of attempts allowed for a frame transmission.
- **Maximum number of node description:** The maximum number of node descriptions received by the network layer before reporting them to the application layer.
- **Maximum response waiting time:** Defines how long a transmitter waits for a response from the receiver.
- **Network scan during:** Determines how long a device spends scanning each channel.
- **NLME-START.Request:** The application layer sends an NLDE-START.Request to the network layer requesting the start of a network. If the network successfully starts, the network layer informs the application layer by sending an NLME-START.Confirm.

11.6 DISCOVERY PROCESS

The originating device uses the discovery command to discover other nodes in the network. The discovery command broadcasts on all three channels and the transmitter switches to receiver mode so that it can receive a response from the nodes.

The network layer supports a discovery process for determining the capabilities of the other nodes in the network. It offers two types of discoveries, NLME-AUTO-Discovery and NLME-Discovery.Request:

- **NLME-AUTO-Discovery:** This primitive is sent from the application layer to the network layer for the purpose of auto-discovery. If the following conditions are met, the destination network layer automatically sends an NLME-DISCOVERY.Confirm to the source:

  1. The LQI of the request command is greater than the LQI threshold in the NIB.
  2. At least one of the device types in the request command matches the destination device type.
  3. At least one profile of the discovery request matches the destination profile.
  4. The auto-discovery request contains the following fields: Destination Capability, Destination Device Type List, Profile ID List, and Auto Discovery Duration.

- **NLME-DISCOVERY.Request:** The originating device uses this command to discover other nodes in the network. This command contains the following fields:

  Destination PAN ID: This is set to 0xffff.

  Destination Network Address: This is set to 0xffff.

  Destination Device Type: The device type to be discovered.

  Source Capability: The capability of the node.

  Source Device Type List: List of device types supported by the source.

  Source Profile Type List: List of profiles supported by the source.

  Discover Profile List Size: The number of profile identifiers in the profile list.

  Discover Profile List: The list of profile identifiers in the discovery request. The discovery response contains only those profile identifiers that match with the profile identifiers in the discovery request.

  Discovery Duration: The maximum MAC symbols a node must wait for response.

If the following conditions are met, the network layer of the destination sends an NLME-DISCOVERY.Indication to the application layer. The application layer decides whether to respond.

- The LQI of the request command is greater than the LQI threshold in the NIB.
- At least one of the device types in the request command matches the destination device type.
- At least one profile of discovery request matches the destination profile.

Discovery Response: When the preceding conditions are met, the application layer of each device will respond to the source with an NLME-DISCOVERY.Response. When the network layer of the source device receives all the responses, it informs the application layer with an NLME-DISCOVERY.Confirm. The NLME-DISCOVERY.Confirm contains the following fields: Status, Number of Discovered Nodes, and Pointers to the Node Description.

Each node description contains the following information: channel, PAN ID, MAC address, node capability, vendor ID, vendor string, device type list, profile list, and request LQI.

Discovery Indication: The discovery indication contains the following fields: Status, Source MAC Address, Source Capability, Vendor ID, Vendor String (Name of the Vendor), Device List Type, Profile ID List, Requested Device Type, and Request LQI.

11.7 PAIRING PROCESS

The pairing process allows two nodes to establish a link between each other to exchange information. The pairing request can be initiated by the controller node or the target node, but the recipient of the pairing request must be a target node. The pairing request is used so that a specific remote control will work only with a specific device. (For example, the TV remote control works only with a specific TV). When the target node receives a command from the controller node, it checks its pairing table. If the controller is not listed in its table, the target node ignores the command. The controller node obtains the PAN ID and MAC address of target node through the discovery process. It then sends an NLME-PAIR.Request to the target. The NLME-PAIR.Request contains the following fields:

- Destination Channel (The two devices will use this channel for their communication.)
- Destination PAN ID
- Destination IEEE Address
- Source Application Capability
- Source Device ID List
- Source Profile List
- Key Exchange Transfer Count (Defines the number of transfers made by the target before changing the link key)

When the target node receives an NLME-PAIR.Request, it makes an entry in its pairing table for the new node.

The pairing table contains the following fields:

- Short Address
- Operating Channel
- IEEE Address
- PAN ID
- Short Address
- Link Key
- User String of New Device (Name of the Node)

When the target node adds a new entry to its table, it responds to the source request with an NLME-PAIR.Response, which contains the following fields:

- Status (of the pairing; e.g., success).
- Destination PAN ID
- Destination IEEE Address
- Source Application Capability
- Source Device Type List
- Source Profile ID List
- Device ID (represents the pairing ID)

When the network layer of the source receives an NLME-PAIR.Response, it issues an NMLE-PAIR.Confirm to the application layer. The NLME-PAIR.Confirm contains following fields:

- Status (e.g., success, aborted)
- Device ID (the address in the pairing table)
- Vendor ID
- Vendor String
- Application Capability

SUMMARY

- The ZigBee RF4CE protocol is used to replace the current IR Remote Control for Consumer Electronics.
- The RF4CE protocol was developed by a consortium that includes Panasonic, Philips, Samsung, and Sony and was recently merged with ZigBee Alliance, becoming ZigBee RF4CE.
- ZigBee RF4CE defines two types of nodes: controller nodes and target nodes.
- ZigBee RF4CE uses a multistar topology.
- The ZigBee RF4CE protocol architecture consists of IEEE 802.15.4 physical, IEEE 802.15.4 MAC, and RF4CE network and application layers.
- ZigBee RF4CE does not require line of sight.
- ZigBee RF4CE operates at the 2.4GHz band.
- ZigBee RF4CE supports frequency agility.
- ZigBee RF4CE uses two-way RF communication between the remote and device.
- ZigBee RF4CE supports pairing schemes.
- Operational channels are 15, 20, and 25.
- In ZigBee RF4CE, only the target node can start a network.
- The pairing command allows two devices to communicate with each other.
- The NLME-DISCOVERY.Request allows a device to discover other nodes in the network.

REFERENCES

1. ZigBee Alliance, www.zigbee.org/rFAQ/tabid/413/Default.aspx

2. Freescale Semiconductor, RF4CE Network Reference Manual

3. Freescale semiconductor, Radio Frequency for Consumer Electronics (RF4CE) Reference Manual, 2/2009

4. Texas Instruments, "Change Infrared to RF with RemoTI," www.ti.com/corp/docs/landing/RF4CE/index.htm

5. Peder Rand, "What's Required for RF4CE?" www.eetasia.com

6. Mindteck, Radio Frequency for Consumer Electronics (RF4CE)

APPENDIX A

6LOWPAN

INTRODUCTION

The Internet Engineering Task Force (IETF) developed the standard for transmitting IPv6 packets over IEEE 802.15.4 low-power wireless personal-area network (6lowpan) for the development of wireless sensor and control networks. Advantages of 6lowpan over other technologies include the interoperability with other wireless networks such as WiFi and WiMax, use of existing protocols such as Transmission Control Protocol (TCP) for end-to-end reliability and Simple Network Management Protocol (SNMP) for management of the network, and it does not require the development of new standards as is the case for other technologies such as ZigBee. Hypertext Transfer Protocol (HTTP), Hypertext Markup Language (HTML), and Extensible Markup Language (XML) can all be used in applications for 6lowpan.

Figure A.1 shows the protocol architecture for 6lowpan. In this architecture, a new layer is added to the current TCP/IP protocol architecture: the adaptation layer.

Due to the growth of the Internet and the address limitations of IPv4, in 1995, the IETF approved IPv6. The limitations that led to Internet Protocol version 6 are summarized here.

The IPv4 address size is 32 bits and can connect up to 2^{32} (4 billion) users to the Internet. The IPv4 address field is divided into two parts, the network address (network ID) and the host address (host ID). After a network number is assigned to an organization, the organization has control of all the host IDs in the host ID field, which it may, or may not, use; this means that all IPv4 addresses might not be used completely. Also, the number of networks connected to the exterior gateway increases rapidly, and this causes the routing table to become large. When the routing table becomes large, it takes more time to search through the table.

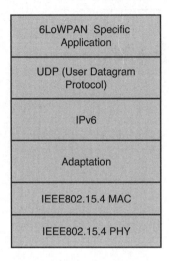

Figure A.1   Protocol architecture of 6lowpan

The IPv6 protocol will reduce the size of the routing table in exterior gateways because IPv6 uses a hierarchical scheme to define an IP address. IPv6 has the following features:

- Expanded addressing
- Simplified header format
- Support extension
- Flow labeling
- Authentication and privacy

A.1 IPV6 STRUCTURE

IPv6 is divided into two parts: a basic header and an extension header. The first 40 bytes of the header is called the basic header, as shown in Figure A.2. IPv6 addresses are 128 bits in length. Addresses are assigned to individual interfaces on the nodes rather than to the nodes themselves. A single interface may have multiple addresses.

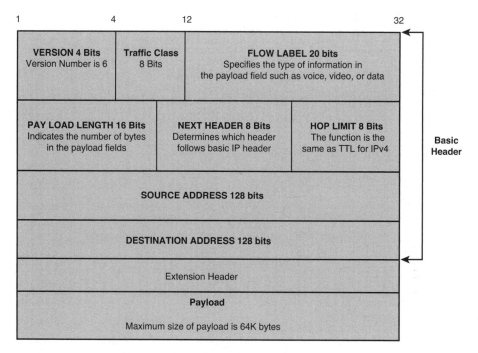

Figure A.2   IPv6 basic header

The fields shown in Figure A.2 are as follows:

- **Version:** The version field is 4 bits; it is 0110 (6 in decimal).
- **Traffic Class:** These 8 bits are used by originating nodes to identify and distinguish between different classes or priorities of IPv6 packets.
- **Priority:** This field defines the priority of an IP datagram. It is used for congestion control. For example, email has a priority of 2, and interactive traffic has a priority of 6.
- **Flow Label:** The 24-bit flow label is used for the labeling of packets that belong to a particular traffic type, such as real-time traffic, that the sender requests for special handling.
- **Payload Length:** 16 bits. Indicates the number of bytes in the payload field.
- **Next Header:** 8 bits. Determines which header follows the basic IP header.
- **Hop Limit:** 8 bits. The function of this field is the same as Time To Live (TTL) for IPv4.

A.2 USER DATAGRAM PROTOCOL (UDP)

User Datagram Protocol (UDP) allows applications to exchange individual packets over a network as datagrams. A UDP packet sends information to the IP protocol for delivery. There is no guaranteed reliability. Figure A.3 shows the UDP packet format.

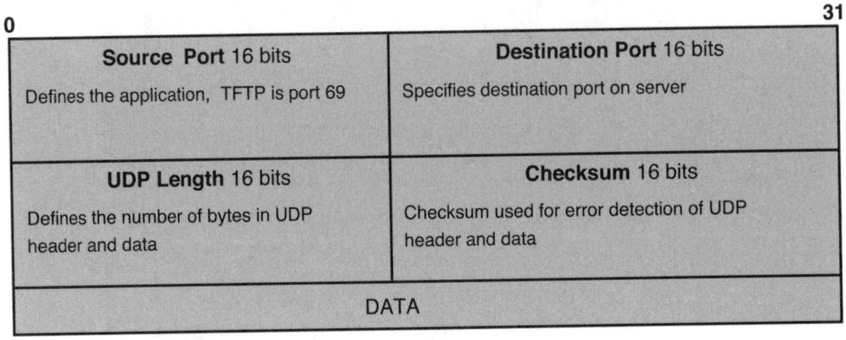

0	31
Source Port 16 bits Defines the application, TFTP is port 69	**Destination Port** 16 bits Specifies destination port on server
UDP Length 16 bits Defines the number of bytes in UDP header and data	**Checksum** 16 bits Checksum used for error detection of UDP header and data
DATA	

Figure A.3   UDP packet format

A.3 IEEE 802.15.4 MAC AND PHYSICAL LAYER FRAME FORMAT

IEEE 802.15.4 developed the MAC and physical layers for low-power wireless personal-area network (LoPWAN). Figure A.4 shows the MAC and physical layer frame formats of IEEE 802.15.4. The maximum size of the physical service data unit (PSDU) is 127 bytes. The size of the auxiliary security field is variable and depends on the size of the message integrity code (MIC): MIC-32, MIC-64, or MIC-128. Assuming the auxiliary security is 8 bytes, the size of the MAC header will be 31 bytes, and the MAC footer will be 2 bytes. The total size of the MAC header and footer will then be 33 bytes. By subtracting the MAC header and MAC footer from 127 bytes, the resulting 94 bytes means the maximum PSDU is 94 bytes. The IPv6 header is 40 bytes, and the UDP header is 8 bytes, so the actual application payload is 56 bytes. Therefore, an application payload of more than 56 bytes cannot be delivered by the physical layer. To overcome this deficiency, the IETF added the adaptation layer between the MAC and IPv6 to adapt the application frame payload to the physical layer payload size.

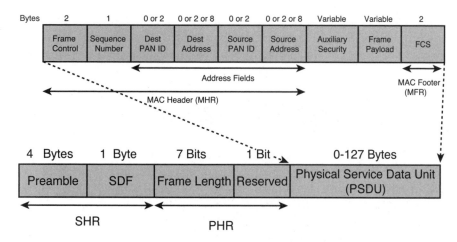

Figure A.4 MAC frame and physical layer frame format of IEEE 802.15.4

A.4 64-BIT GLOBAL IDENTIFIER

The IEEE defines a 64-bit extended unique identifier (EUI-64), which is called the EU. It is divided into two parts: organization unique identifier (OUI) and the interface identifier. Figure A.5 shows the EUI-64.

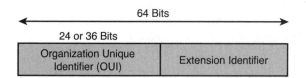

Figure A5 EUI-64 addressing format

The OUI represents the company ID; it can be 24 bits or 36 bits. This field is assigned by the Institute of Electrical and Electronic Engineers (IEEE) Registration Authority, and the extension identification is assigned by the company.

A.5 ADAPTATION LAYER

The function of adaptation layer is to perform following tasks:

- Compress the IPv6 header
- Fragment the IPv6 payload
- Compress the UDP header

The adaptation layer is positioned between the MAC and IPv6 layers. It adds 8 bits to the frame, which defines if the IPv6 header is uncompressed or compressed. Figure A.6 shows the general frame format of 6lowpan at the MAC level, which includes the Adaptation field.

Figure A.6 MAC frame format with adaptation header

There are four different types of adaptation headers, as shown in Figure A.7, which are indicated by the first 2 bits of the Adaptation Header field:

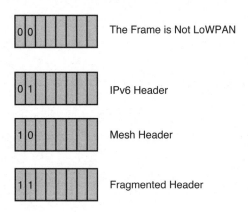

Figure A.7 Adaptation headers

- **Adaptation header for IPv6 header:** The adaptation header for the IPv6 header is 01XXXXXX, which is called the dispatch byte. If the first 2 bits of the adaptation layer are 01, the other bits determine the types of IPv6 header. **01**000001 means the IPv6 header is uncompressed, whereas **01**000010 means it is compressed. Figure A.8 shows the adaptation header with an uncompressed network header.

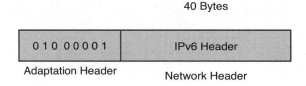

Figure A.8 Uncompressed header

- **IPv** ... ith a value of 01000010 indicates
  that ... er Compression One (HC1) field
  dete ... ressed. It also determines whether
  ther ... e fields in the IPv6 header can be
  con ... A.9 shows the compressed IPv6
  hea ... Pv6 header are compressed into 2
  byt

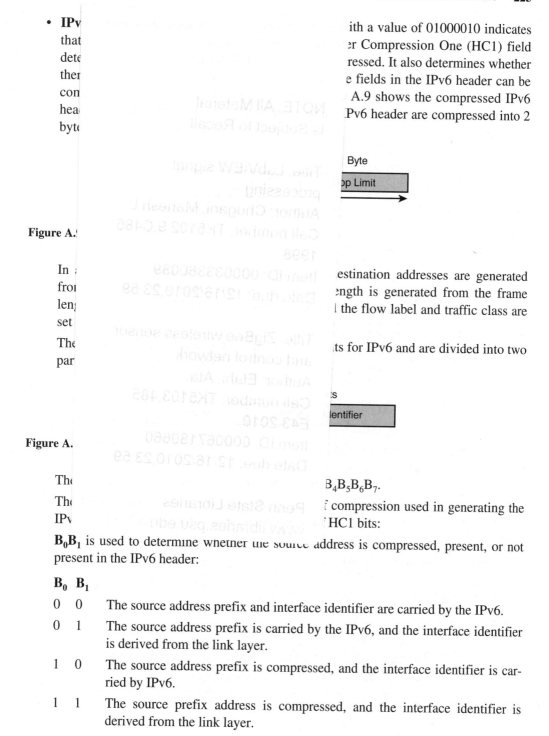

Byte

op Limit

Figure A.

In ... estination addresses are generated
fro ... ength is generated from the frame
len ... the flow label and traffic class are
set

The ... ts for IPv6 and are divided into two
par

dentifier

Figure A.

Th ... $B_4B_5B_6B_7$.

Th ... compression used in generating the
IPv ... HC1 bits:

B_0B_1 is used to determine whether the source address is compressed, present, or not present in the IPv6 header:

B_0 B_1

0 0   The source address prefix and interface identifier are carried by the IPv6.

0 1   The source address prefix is carried by the IPv6, and the interface identifier is derived from the link layer.

1 0   The source address prefix is compressed, and the interface identifier is carried by IPv6.

1 1   The source prefix address is compressed, and the interface identifier is derived from the link layer.

B_2B_3 is used to determine whether the destination address is compressed, present, or not present in the IP header

B_2 B_3

0 0   The destination address prefix and interface identifier are carried by the IPv6.

0 1   The destination address prefix is carried by the IPv6, and the interface identifier is derived from the link layer.

1 0   The destination address prefix is compressed, and the interface identifier is carried by the IPv6.

1 1   The destination address prefix is compressed, and the interface identifier is derived from the link layer.

B_4 is used to define if 8-bit traffic class and 20-bit of flow label are included in the IPv6 header:

B_4

0   The 8-bit traffic class and 20-bit flow label are included in the IPv6 header.

1   The traffic class and flow label are compressed (not included in the IPv6 header).

B_5B_6 defines the next header after adaptation header:

B_5 B_6

0 0   The next header is IPv6 and it is not compressed.

0 1   The next header is UDP.

1 0   The next header is ICMP.

1 1   The next header is TCP.

B_7 is used to indicate whether there is more HC header:

B_7

0   No more HC header.

1   More header compression (HC) follows HC1.

When B_5B_6 of the HC1 is set to 01 and B_7 is set to 0, next header is UDP, as shown in Figure A.11.

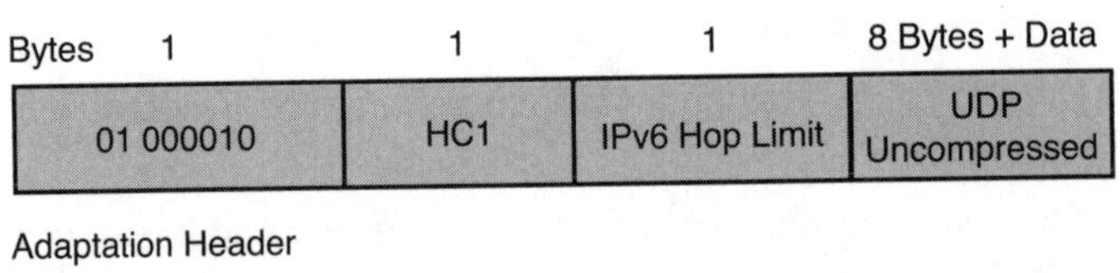

Bytes 1	1	1	8 Bytes + Data
01 000010	HC1	IPv6 Hop Limit	UDP Uncompressed

Adaptation Header

Figure A.11   HC with uncompressed UDP

- **HC-UDP compressed format:** When the B_5B_6 bits are set to 01, the next header is UDP; and when B_7 is set to 1, HC-UDP follows HC1, and the UDP header is compressed, as shown in Figure A.12.

01000010	HC1 –IPv6	HC2(HC-UDP)	IPv6 Header Compressed	UDP Header Compressed

Figure A12   IP header and UDP header compressed

HC2-UDP is represented by 8 bits: $B_0 \, B_1 \, B_2 \, B_3 \, B_4 \, B_5 \, B_6 \, B_7$. It determines which fields of the UDP header are compressed:

$B_0 = 0$ UDP source port is not compressed, and it is carried by the UDP header.

$B_0 = 1$ UDP source port is compressed to 4 bits; the actual decimal value of the port number can be determined by adding 61616 to the 4 bits.

The UDP ports assigned for 6lowpan are between 61616 and 61631. Therefore, the source and destination port can be compressed in UDP header as 4 bits:

$B_1 = 0$ UDP destination port is not compressed.

$B_1 = 1$ UDP destination port is compressed.

$B_2 = 0$ UDP length is not compressed and it is carried within the UDP header.

$B_2 = 1$ UDP length is compressed and is calculated as follows.

UDP length = IPv6 payload length minus the length of extension header minus the UDP header.

Bits B_3 through B_7 are reserved.

- **Adaptation headers for mesh network:** Figure A.13 shows the mesh addressing format that is used in networks with mesh topologies.

1	0	V	F	Hops Left	Originator Address 16/64	Final Address 16/64

Figure A.13   Adaptation header for mesh network

In Figure A.13, the originator address and the final address are the link layer addresses:

- **Originator Address:** Link layer address of the originating packet.
- **Final Address:** Link layer address of the destination.
- **V:** Set to zero if the source uses a 64-bit address, and set to one if the source of the packet uses a 16-bit address.

- **F:** Set to zero to indicate that the final destination uses a 16-bit (short) address. Otherwise, when it is set to one, it indicates that the final destination uses a 64-bit address.

- **Hops Left:** The maximum number of nodes a packet can travel to reach the destination. This field is decremented by one before the packet is transmitted to the next node. When this field becomes zero, the packet will not be forwarded to the next node. If the value of Hops Left is set to 0xF (1111), there is another 8-bit extension field that follows the 4 bits.

Figure A.14 shows the mesh header with compressed IPv6 and UDP headers.

1 0 V F Hops Left	01000010	HC1 for IPv6	HC2 for UDP	IPv6 Header Compressed	UDP Header Compressed

Figure A.14 Mesh header with IPv6 compressed and UDP compressed

A.6 FRAGMENTED IPV6 PAYLOAD

The IPv6 payload does not fit in the IEEE 802.15.4 physical layer payload and therefore must be fragmented. Figure A.15 shows the first header of a fragmented payload, and Figure A.16 shows the subsequent header.

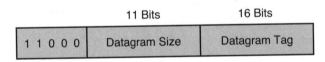

	11 Bits	16 Bits
1 1 0 0 0	Datagram Size	Datagram Tag

Figure A.15 First fragmented header for IPv6

- **Datagram Size:** This 11-bit field represents the number of bytes in an IPv6 payload.

- **Datagram Tag:** This number is selected by the originator of the frame and is incremented with the sending of each fragment of the frame.

- **Datagram Offset:** This 8-bit field represents where this fragment is located within the entire frame; each offset unit represents 8 bytes.

	11 Bits	16 Bits	8 Bits
1 1 1 0 0	Datagram Size	Datagram Tag	Datagram Offset

Figure A.16 Next fragmented header for IPv6

APPENDIX B

WIRELESS HART

INTRODUCTION

HART is an acronym that stands for Highway Addressable Remote Transducer. It is a wired control and sensor network that is used for process control, environmental monitoring, energy management, plant monitoring, advanced diagnostics, plant automation, and equipment calibration. Currently, 22 million HART devices are in use (and were installed by the HART organization). Due to the demand for wireless sensor networks, the HART organization published new specifications for wireless HART (HART Protocol Revision 7) that are compatible with wired HART. Figure B.1 shows the Wireless HART protocol architecture and the Open Systems Interconnection (OSI) reference model.

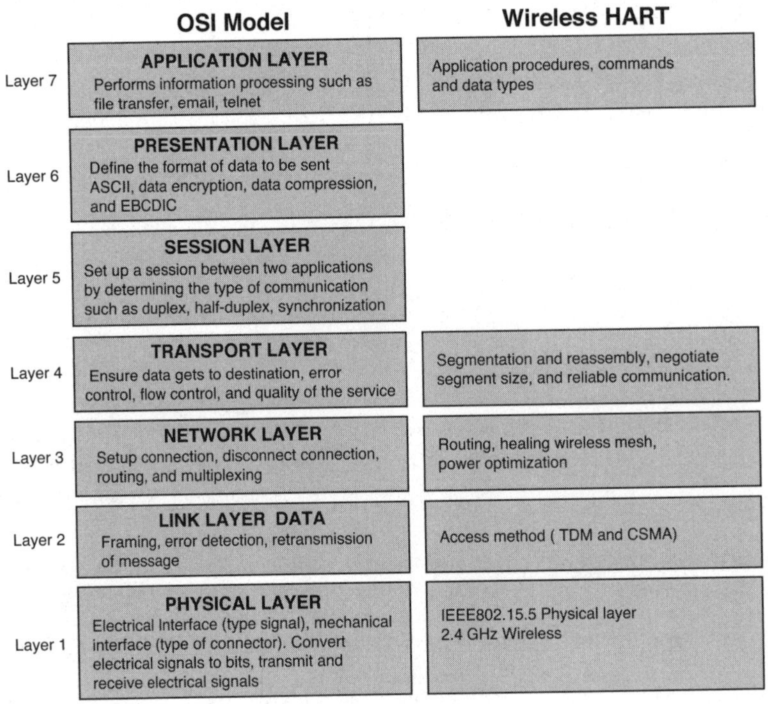

Figure B.1 Wireless HART protocol architecture

B.1 WIRELESS HART PHYSICAL LAYER

The physical layer of the wireless HART protocol is based on the IEEE 802.15.4 standard. The physical layer operates at the 2.4GHz band using direct-sequence spread spectrum (DSSS) and offset quadrature phase-shift keying (O-QPSK) modulation with frequency-hopping spread spectrum (FHSS). The data rate of wireless HART is 250Kbps (62.5K signals per second).

The 2.4GHz band offers 16 channels. The wireless HART device can use any hopping pattern and block those channels that are causing interference. It requires only a nominal transmitter power (+10 dBm), and its maximum transmission range is 200 meters in free air.

B.2 WIRELESS HART DATA LINK LAYER

Wireless HART offers time-division multiple access (TDMA) and carrier-sense multiple access (CSMA) for its access methods. In the TDMA method, multiple users access the network using a single frequency without interference. This is done by allocating time slots to each device. Each device must keep track of the time and synchronize its time with its neighbor. A device communicates with other devices based on prescheduled time slots that were allocated to the device. Because each device has its own time slots, it is collision free. TDMA scheduling is a series of 10ms time slots, where each device is able to use the network for 10ms. The MAC layer assigns a 64-bit (long) address or a 16-bit (short) address to each device.

The HART device supports multiple super frames with different time slots for different types of traffic (for example, fast or slow transactions). For reliable communication between nodes, the MAC layer uses acknowledgments and assigns each packet an ID number.

B.3 WIRELESS HART NETWORK LAYER

The network layer creates the topology for the network: mesh, star, or star-mesh. In a mesh network, all devices in the network are full-function devices. That is, they all support routing. Wireless HART supports full-mesh topology, where every device can have multiple routes.

The network layer can perform self-healing based on the condition of the device and on the environment in which the device is located. If the path between two devices is blocked due to an obstacle, the network layer can find another route automatically so that these two devices can communicate with each other.

A wireless HART device is able to discover its neighbor and get synchronization information (time and frequency-hopping patterns) from its neighbor nodes. It can then establish a link to its neighbor. Besides receiving and transmitting data, a device can also relay information from other devices for routing.

The network layer supports multicast, broadcast, and unicast transmissions and message routing.

B.4 WIRELESS HART NETWORK COMPONENTS

The wireless HART network components are the wireless HART network manager, the wireless HART field device, the wireless HART gateway, and the wireless HART adapter. Figure B.2 shows wireless HART network components.

- **Wireless HART network manager:** Controls the mesh topology and devices in the network. Each network can have only one wireless manager, which performs the following functions:
  - Network formation
  - Permits devices to associate and disassociate from the network
  - Sets communication time slots for the devices in the network
  - Finds the route from the source to the destination

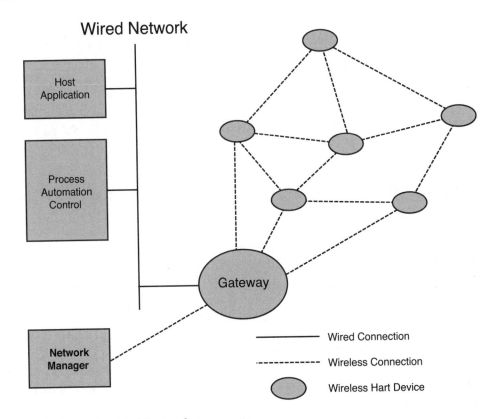

Figure B.2 Wireless HART network components

- Monitors devices in the network
- Communicates with the security manager to obtain the security key and authentication information
- **Wireless HART field device (WFD):** Connects to the equipment of the plant for receiving and transmitting data. It can also accept routing data and pass that information to the next node.
- **Wireless HART gateway:** Connects a wireless HART network to a wired network.
- **Wireless HART adapter:** Connects wired HART devices to wireless networks.

B.5 NETWORK FORMATION

The following commands are used to form a wireless HART network.

- **Advertise:** The devices in a network send the advertising command to the network manager to inform it of their existence.

- **Join:** A new device sends a join request command to the network manager. The network manger authenticates the new device and responds with an activation packet informing the new device that it has been accepted or rejected.

- **Schedule:** The network manager requests information from a device, such as how often the device transmits data or the bandwidth required by the device. The network manager then uses this information to send the transmission schedule to the new device and its intermediate router.

B.6 SECURITY

Wireless HART offers robust security for protecting data and the network via the following security elements:

- **Confidentiality:** Apply end-to-end data encryption by using the Advanced Encryption Standard (AES-128). All traffic from the sensor to the network is encrypted; this includes both the data and the commands.

- **Verification:** Message integrity code (MIC) is generated for all data.

- **Anti-jamming:** This is achieved by changing the hopping pattern.

- **Authentication:** Any new device joining the network must be authenticated by the network manager. The network manager also monitors unauthorized devices that try to join the network.

- **Message integrity:** Message integrity and security codes are offered to prevent the duplication of a message.

- **Key management:** Wireless HART uses multiple keys for security, as follows:

  - *Device key:* Each device has it own key, which it uses for joining the network. Only trusted devices identified by their key, device ID, and manufacturer ID are allowed to join the network.

  - *Session key:* Used for device authentication.

  - *Network key:* Used for encryption of the network payload.

B.7 WIRELESS HART DATA TRANSFER MODE

The data types depend on the device application and how often the data is collected from the application. Some of the data types are as follows:

- **Periodic sampling:** This is used for applications that require continuous monitoring, such as the temperature of device or pressure of a pipeline. The application periodically sends the data.

- **Event driven:** The sensor monitors events (changes) such as a fire alarm or the opening and closing of a door or window. The sensor will transmit the data when the event occurs.

- **Store and forward:** In some applications, data is not so critical that it needs to be transferred right away. The data can be stored in the sensor node and transmitted based on a schedule.

- **Time-triggered action:** At a specific time, the device transmits the information.

- **Time-stamped data:** The time that the data was taken from the device is transmitted with the data.

Table B.1   Characteristics of Wireless HART

Characteristic	Description
Frequency band	2.4MHz ISM band
Data rate	250Kbps
MAC and physical layers	IEEE802.15.4
Transmission distance	200 meters (line of sight [LOS])
Power	Battery, line power, solar
Topology	Star, mesh, star-mesh
Number of devices	No limit

APPENDIX C

Z-WAVE

INTRODUCTION

Z-Wave was developed by the Z-Wave Alliance to control lights, thermostats, meter readings, home entertainment, and sensors. The primary application of the Z-Wave technology is for wireless home control, such as light switches, thermostats, blind/drapes, and security. Z-Wave represents an alliance of more than 160 manufacturers that created a standard for Z-Wave (including Intel, Intermatic, Leviton, and Universal Electronics).

Z-Wave uses mesh topology, which can self-organize and self-heal. If communication between two nodes fails because of an obstacle, Z-Wave can route the message to the destination through other nodes (robust routing). A Z-Wave network can have a maximum of 232 nodes and uses short messages to communicate between nodes.

C.1 Z-WAVE PROTOCOL ARCHITECTURE

Figure C.1 shows the Z-Wave protocol architecture. As shown, the Z-Wave protocol architecture is made of five layers: application, routing, transfer, MAC, and RF Media.

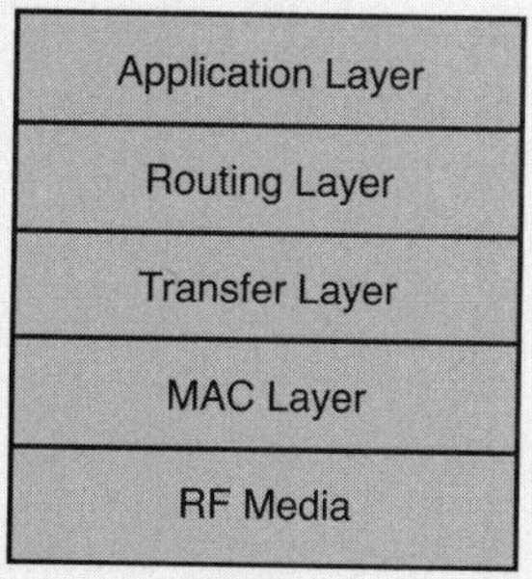

| Application Layer |
| Routing Layer |
| Transfer Layer |
| MAC Layer |
| RF Media |

Figure C.1 Z-Wave protocol architecture

C.2 RF MEDIA

Z-Wave operates in the 908MHz band of the Industrial, Scientific, and Medical (ISM) band in the United States, and in the 860MHz band in Europe. In the European countries, Z-Wave's physical layer operates at 868.42MHz and has a duty cycle of 1%. This means that only 1% of the time is the device allowed to transmit. Z-Wave uses Manchester encoding: The clock is embedded with the data signal. In Manchester encoding, there is a transition from zero to one in the middle of zero, and there is a transition from one to zero in the middle of one. The Manchester encoding is transmitted using frequency-shift keying (FSK) modulation.

C.3 MAC LAYER

The Z-Wave Media Access Control (MAC) layer performs the following functions.

- Accepts signals from the radio frequency (RF) media and makes the frame
- Controls the RF media layer but is independent of the RF media
- Uses carrier-sense multiple access with collision avoidance (CSMA/CA) for its access method to prevent collision. When the media is busy, the MAC layer randomly delays its transmission.

Figure C.2 shows the Z-Wave MAC frame format.

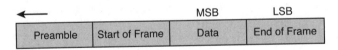

Figure C.2   Z-Wave MAC layer frame format

APPENDIX D

ABBREVIATIONS

6lowpan	IPv6 over low-power personal-area network
ACK	acknowledgment
ACL	access control list
ADOV	Ad hoc On-Demand Distance Vector
AES	Advance Encryption Standard
AIB	APS Information Base
AMA	active member address
AMI	advance metering infrastructure
APDU	application protocol data unit
APS	application support sublayer
APSDE	application support sublayer data entity
APSME	application support sublayer management entity
ASDU	application support data unit
ASK	amplitude-shift keying
BCC	block check character
BI	beacon interval
BLE	battery life extension
BO	beacon order
BPSK	binary phase-shift keying
BS	base station
BSN	beacon sequence number
BWA	broadband wireless access
CAP	contention access period
CBC	cipher block chaining
CBKE	certificate-based key establishment
CCA	clear channel assessment
CCK	complementary code keying

CCM	cipher block chaining message authentication code
CRC	cyclic redundancy check
CS	convergence sublayer
CSMA/CA	carrier-sense multiple access with collision avoidance
CSMA/CD	carrier-sense multiple access with collision detection
CTR	counter
CW	contention window
DES	Data Encryption Standard
DSN	data sequence number
DSR	Dynamic Source Routing
DSSS	direct-sequence spread spectrum
EC	elliptic curve
ED	energy detection
EP	endpoint
ESP	energy service portal
FCS	frame check sequence
FFD	full-function device
FHSS	frequency-hopping spread spectrum
FSK	frequency-shift keying
FTP	File Transfer Protocol
GTS	guaranteed time slot
HA	home automation
HAN	home-area network
HCI	host controller interface
HTML	Hypertext Markup language
HTTP	Hypertext Transfer Protocol
HVAC	heating, ventilation, and air conditioning
IAS	intruder alarm system
ICMP	Internet Control Message Protocol
IEEE	Institute of Electrical and Electronics Engineers
IETF	Internet Engineering Task Force
IFS	interframe spacing
IP	Internet Protocol
IPM	industrial plant monitoring
IR	infrared
ISM	Industrial, Scientific, and Medical band
LLC	logical link control
LMP	Link Management Protocol
LQI	link quality indication
LR-WPAN	low-rate wireless-personal network
MAC	Media Access Control
MCPS	MAC common part sublayer

MIB	MAC layer Information Base
MIC	message integrity code
MIMO	multiple input, multiple output
MLME	MAC (layer) management entity
MPDU	MAC protocol data unit
NAN	neighbor-area network
NIB	Network (layer) information base
NIC	network interface card
NLDE	network layer data entity
NLME	network layer management entity
NLOS	no line of sight
NPDU	network layer protocol data unit
NSDU	network service data unit
OFDM	orthogonal frequency-division multiplexing
O-QPSK	offset quadrature phase-shift keying
PAN	personal area network
PAN ID	personal area network identification
PD-SAP	physical data service access point
PDU	protocol data unit
PHHC	personal, home, and hospital care
PIB	Physical Layer Information Base
PLME	physical layer management entity
PMP	point to multipoint
POS	personal operating space
PPDU	physical layer protocol data unit
PSK	phase-shift keying
PSSS	parallel-sequence spread spectrum
PTC	programmable communication thermostat
QAM	quadrature amplitude modulation
QoS	quality of service
QPSK	quadrature phase-shift keying
RF	radio frequency
RF4CE	Radio Frequency for Consumer Electronics
RFD	reduced-function device
RFID	radio frequency identification
RREP	route reply
RREQ	route request
RSSI	received signal strength indicator
Rx	receiver
SAP	Service Access Point
SD	superframe duration
SE	smart energy

SFD	start frame delimiter
SKKE	Symmetric-Key Key Establishment
SNR	signal-to-noise ratio
SS	subscriber station
TA	telecom application
TCP	Transmission Control Protocol
Tx	transmitter
UDP	User Datagram Protocol
UNII	Unlicensed National Information Infrastructure Band
WiFi	wireless fidelity
WLAN	Wireless LAN
WPAN	wireless personal-area network
WSN	wireless sensor network
ZDO	ZigBee device object
ZDP	ZigBee device profile
ZCL	ZigBee Cluster Library
ZTC	ZigBee trust center

BIBLIOGRAPHY

"Advance Metering Infrastructure (AMI): Smart Energy," www.aeri.net

Alpha Systems, Inc., Japan, "DLNA/UPnP-ZigBee Gateway Specifications"

Ashton, S., Ember Corp, "Designing Smart Energy Devices"

Atmel Corp., "Low Power 2.4 Transceiver for ZigBee, IEEE 802.15.4, and ISM Applications," www.atmel.com/dyn/resources/prod_documents/doc8111.pdf

Cirticom Corp, Securing Sensor Network, http://www.certicom.com, March 2006

Culler, D. E. and Hui, J. "IP on IEEE802.15.4 Low-Power Wireless Network," Arch Rock Corporation, 2007

Cunha, A., Alves, M., and Koubàa, A. "Implementation Details of the Time Division Beacon Scheduling Approach for ZigBee Cluster-Tree Networks," www.hurray.isep.ipp.pt, 2007

Daintree Networks, "Building and Operating Robust and Reliable ZigBee Networks"www.daintree.net/downloads/.../robust_zigbee_network.pdf

Daintree Networks, "Comparing ZigBee Specification Versions," www.daintree.net/resources/spec-matrix.php

Daintree Networks, "Understanding ZigBee Commissioning"

Daintree Networks, "ZigBee Specification Update," www.zigbee.org/ZigBeeSpecificationDownloadRequest/.../Default.aspx

Document 053474r17, "ZigBee Specification," ZigBee Alliance, October 2007

Document 053520r25, "ZB HA PTG Home Automation Profile," ZigBee Alliance, October 2007

Document 064321r09, "ZigBee Stack Profile," January 2008

Document 074855r05, "ZigBee PRO Stack Profile," ZigBee Alliance, January 2008

Document 075123r01ZB, "ZigBee Cluster Library" ZigBee Alliance, Oct 2007

Document 075123r02, "AFG-ZigBee Cluster Library Specification," ZigBee Alliance, May 2008

"Elliptic Curve Cryptography," http://en.wikipedia.org

Ember Corp, "ZigBee Security," http://portal.ember.com/node/685

Ember Corp, "ZNet Application Developer's Guide," July 2006

"Estimate Transmission Range for ZigBee and Propriety Short-Range Wireless Devices in 900MHz and 2.4GHz Band," www.ednasia.com

Freescale Semiconductor, "ZigBee Cluster Library Reference Manual Rv1.1," Oct 2008

Freescale Software Reference Manual for ZigBee 2007, Freescale Semiconductor Corp, Jan 2008

Grossmann, R., and Blumenthal, J., *Localization in ZigBee-Based Sensor Network*, University of Rostock, www.loms-itea.org/publications/LocalizationZigbee.pdf

Hart Technology, www.hartcomm2.org

"IEEE Standards EUI-64 Tutorial", www.standards.ieee.org/regauth/oui/tutorials/UseOfEUI.html

Jennie Corp., "ZigBee Application Framework API," Reference Manual JN-RM-2018, Revision 1.5, 2007

Kinney, P. "Gateways: Beyond the Sensor Network," ZigBee.org

Nieuwenhuyse, A., and Koubàa, M. "The Use of the ZigBee Protocol for Wireless Sensor Networks," HURRAY-TR-060603 2006

Pan, M. "The Orphan Problem in ZigBee-based Wireless Sensor Networks," Department of Computer Science, National Chiao Tung University

Pan, M., and Tseng, Y. *ZigBee Wireless Sensor Networks and Their Applications,* Department of Computer Science, National Chiao Tung University

RFC 4292, *IPv6 Addressing Architecture*

RFC 4944, *IPv6 over 802.15.4*

Steigmann, R. "Introduction to WISA," ABB STOTZ-KONTAKT GmbH, Germany

Sturek, D. "ZigBee Technical Overview," www.Zigbee.org

TSMP Technology, www.dustnetworks.com/technology/tsmp.shtml

Wang, R., Chang, R., and Chao, H. "Internetworking between ZigBee/802.15.4 and IPv6/802.3 Network," Department of Computer Science and Information Engineering, National Dong Hwa University, Taiwan, R.O.C.

"Wireless Infrastructure for Your Application," www.alektrona.com

Yum, Beun, Lee "Method to use 6LoWPAN in IPv4 Network," ieeexplore.ieee.org/xpls/abs_all.jsp?arnumber=4195321

Z-wave Technology, www.z-wave.com

INDEX

ZigBee Wireless Sensor and Control Network

Ata Elahi with Adam Gschwender

Prentice Hall Communications Engineering and Emerging Technologies Series
Theodore S. Rappaport, Series Editor

FREE Online Edition

FEB – 3 2010

Your purchase of *ZigBee Wireless Sensor and Control Network* includes access to a free online edition for 45 days through the Safari Books Online subscription service. Nearly every Prentice Hall book is available online through Safari Books Online, along with more than 5,000 other technical books and videos from publishers such as Addison-Wesley Professional, Cisco Press, Exam Cram, IBM Press, O'Reilly, Que, and Sams.

SAFARI BOOKS ONLINE allows you to search for a specific answer, cut and paste code, download chapters, and stay current with emerging technologies.

Activate your FREE Online Edition at
www.informit.com/safarifree

> **STEP 1:** Enter the coupon code: ADGAXWA.

> **STEP 2:** New Safari users, complete the brief registration form.
> Safari subscribers, just log in.

If you have difficulty registering on Safari or accessing the online edition, please e-mail customer-service@safaribooksonline.com

 Addison Wesley
 AdobePress
 ALPHA
 Cisco Press
 FT Press
 IBM Press
 lynda.com
 Microsoft Press
 New Riders

 O'REILLY
 Peachpit Press
 QUE
 Redbooks
 SAMS
 SAS Publishing
Sun microsystems
Wiley Publishing
WILEY